Chanticleer Press, New York

Scientific consultant: John E. McCosker
Steinhart Aquarium, San Francisco

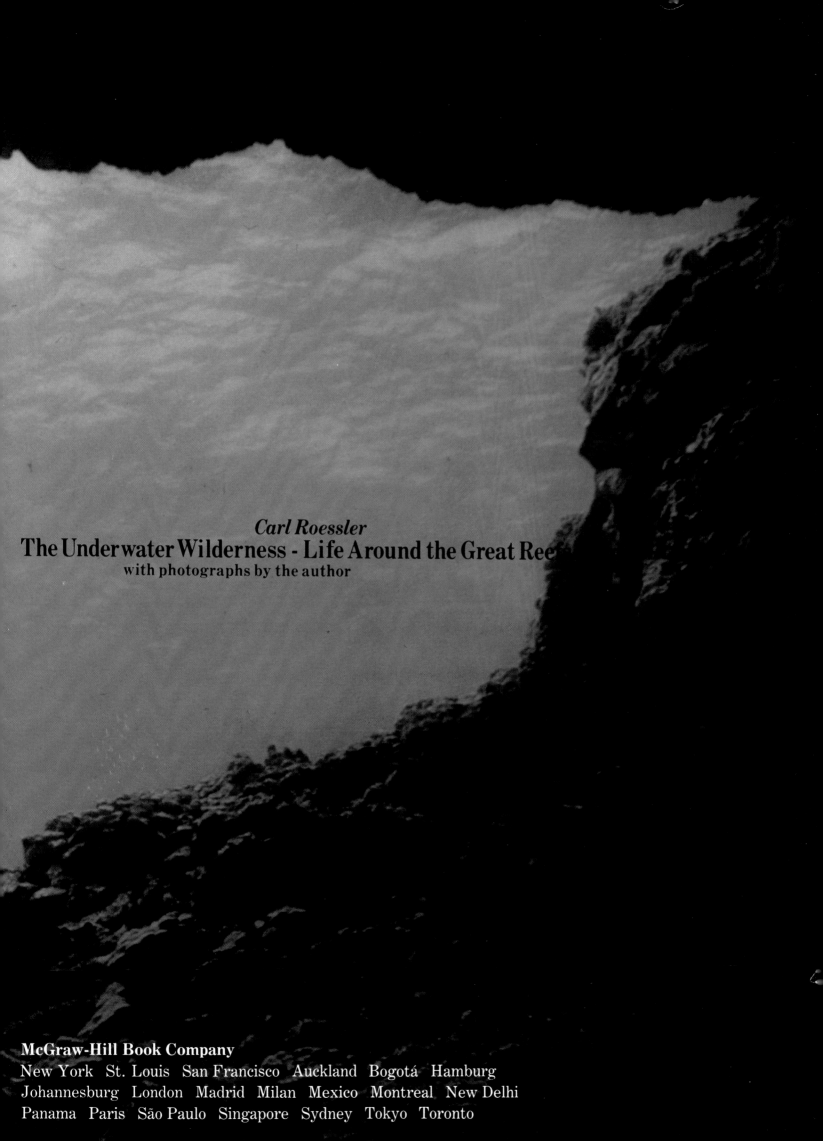

Carl Roessler
The Underwater Wilderness - Life Around the Great Reef
with photographs by the author

McGraw-Hill Book Company
New York St. Louis San Francisco Auckland Bogotá Hamburg
Johannesburg London Madrid Milan Mexico Montreal New Delhi
Panama Paris São Paulo Singapore Sydney Tokyo Toronto

1. Colonies of bushy gorgonians growing at a depth of 35 meters. A diver, entering their silent realm, soars above them. (Cayman Islands, Caribbean)
2. A diver approaching three masked butterflyfish (Chaetodon lunula) on a placid reef. (Rangiroa, French Polynesia)
3. A diver approaching a group of tube sponges in a densely populated section of reef. These sponges are growing at the edge of a precipice thousands of meters deep. (North Wall, Grand Cayman, Caribbean)
4. A diver soaring past the mouth of a cave formed by the interaction of volcanic lava and sea water. Such caves are common in the shallow waters of the Hawaiian Islands. (Kona Coast, Hawaii)
5. A young diver approaching well-developed coral heads. These coral formations nest on a sandy shelf some seven meters deep and rise to within three meters of the surface. (Bonaire, Caribbean)

First published by Chanticleer Press, Inc.,
424 Madison Avenue, New York, N.Y. 10017
Second printing by McGraw-Hill Book Company,
1221 Avenue of the Americas, New York,
N.Y. 10020

ISBN: 0-07-053503-5

Library of Congress
Cataloging-in-Publication Data
Roessler, Carl, 1933–
The underwater wilderness.
Includes index.
1. Coral reef biology. I. Title.
QH95.8.R64   1986   574.9'1   86-27841
ISBN 0-07-053503-5

Printed and bound by Dai Nippon Printing Co., Ltd., Tokyo, Japan

Chanticleer Staff
Publisher: Paul Steiner
Editor-in-Chief: Gudrun Buettner
Executive Editor: Susan Costello
Managing Editor: Jane Opper
Editor: Eva Neumann
Production: Helga Lose
Drawings: Howard S. Friedman

Design: Massimo Vignelli

Acknowledgement
All the photographs in this book are by Carl Roessler with the exceptions of the cover by Chris Newbert/ Bruce Coleman, Inc., and illustrations 281, 282, 285, 286, and 288 by Alain Schweigert, and 283, 284, and 287 by F. Candela Ory.

The author wishes to thank the Sea Library of Los Angeles for the use of some of his photographs from their files.

To Jessica

Contents

Foreword

It is a common saying among the oceanographic community that much of the surface of the moon is better charted than the bottom of the sea. The information collected by ocean-going and scuba-diving researchers in the last 25 years has begun to turn the tide in man's understanding of the forces and inhabitants of a medium that covers nearly 70 percent of the planet earth. Yet we remain at a threshold, and it can be said without reservation that the last, great remaining wilderness area of our planet is the sea. Carl Roessler, who has spent more than two decades either below water exploring or at the surface planning his next descent, is a pioneer in the search for aquatic wilderness areas. His photographs and prose are well known to a devoted following within the diving community, as well as the scientific community which is often the beneficiary of his discoveries. Marine biologists at the California Academy of Sciences are often pleased to discover through his photography a range extension of a rare species, intrigued by his novel observations of animal behavior, such as territorial sharks or sympathetic sea snakes, or baffled when shown a Roessler photograph of a feathery crinoid or a delicate *Dendronephthya* coral which may be new to science.

The fact that a scientific name can be applied to the vast majority of birds or mammals that appear in a book such as this attests to the state of the aquatic art. Scientists are far from being finished with their task of compiling and classifying the near-shore inhabitants and even further from that goal for ocean life forms as deeper depths are explored. It will require more than a decade before all the animals which Carl Roessler has herein portrayed receive their taxonomic names—and certainly a longer period before their biology is understood.

I suspect that after pondering the beauty and stories in this volume, even the most landlocked of readers will become aware of the interconnectedness of the oceans of the world. With this knowledge, the paradox of an oceanic wilderness in a continuous environment will become evident: How can a pristine bay or unspoiled cay exist if it is surrounded by a fluid that is continuous from continent to continent and pole to pole? Will the anti-wilderness ethic of our progress-oriented civilization be responsible for the death of the underwater wilderness? I suspect not—particularly after one gains a sensitivity to and understanding of what is at stake. Carl Roessler has stated his case well.

Dr. John E. McCosker
Director,
Steinhart Aquarium, San Francisco

Introduction

Out of a cloudless equatorial sky the sun pours down on a sparkling aquamarine sea. A deserted sandy beach glistens hot and white against a backdrop of green-fronded palm trees rustling in a light breeze. Beneath the sun-glinting surface of the water the sunlight seems as clear as air. Intricately structured coral colonies are scattered across the bright white-sand bottom as far as the eye can see. Brilliantly colored reef fish hover, dart through the maze, or dance endlessly in the sunlight above their coral shelters. A bit farther from shore, the reef shallows end in an awesome precipice, dropping off abruptly into deep, still, cobalt-blue waters. In the depths drift great schools of jacks and tuna, while nearer the precipice majestic rays, turtles, and sharks make their tireless rounds.

All is life, and light, and motion . . . a tropical island paradise far out on a peaceful sea . . . the underwater wilderness tapping some wellspring of tranquillity deep within us, satisfying a yearning for some pristine world.

But words can hardly do justice to the wonder and beauty of the world under the sea. We need a finer understanding of this "underwater wilderness"—how it works and why it is important to us—and a greater awareness of the dangers that threaten it. The term "underwater wilderness" was chosen for this book to evoke the familiar image of a place where, through isolation or act of law, a natural community of plant and animal life flourishes without man's presence. Significantly, we define a wilderness as a place into which human beings have not intruded. Some of the sites we will describe are protected today only by their remoteness from human settlement. Others border lightly settled areas but have not yet been damaged irreparably by the fishing, waste-disposal, and building development going on nearby. For the moment, then, these enclaves constitute our ideal of underwater wilderness, and we shall so celebrate them.

There are two important reasons for us to understand the concept of an underwater wilderness. By appreciating underwater sites in all their richness and vitality, we can best determine how to protect and cherish them. We can then see clearly how the spread of civilization across the globe threatens many of these magnificent communities with extinction. Viewing the sea in this new way, as a wilderness, and acquiring a broad appreciation of such underwater areas for their natural beauty and value, we can better cope with the forces of destruction that threaten them. There is ample precedent ashore.

As society has spread concrete and macadam across the face of the earth on a colossal scale, it has very reluctantly begun to heed the advice of a minority of people who have urged the preservation of certain portions of the land in their natural state. The setting aside of such areas is an attempt both to husband resources and to create refuges where we and our progeny can enjoy the comparative tranquillity of nature. It has now been widely accepted that most of us need access to open, unspoiled areas for our individual well-being. While the preservation of large areas such as national parks and wilderness areas has been common for decades on all continents, the idea of protecting tropical reefs and sea life is relatively new.

Aside from the psychological value of access to large undeveloped tracts both on land and in the sea, there is an even more pressing reason for preserving underwater wildernesses and their surrounding seas. We must first realize that the oceans are overwhelmingly large and yet pathetically fragile. Ocean waters cover 71 percent of the earth's surface, and their volume is estimated at nearly 700 million cubic kilometers. Our oceans are the source of nearly 80 percent of the planet's oxygen supply, due to the photosynthetic activity of the phytoplankton that live in the upper waters of the sea. The oceans also provide an increasing portion of mankind's protein needs. Unfortunately, for generations we have taken the sea's largesse without regard to the possible effects on its health. We have assaulted the sea with our pollution and harvested its bounty without concern either for its limits or its balance. Only now, as we learn more of the sea's ecosystem, do we realize how vulnerable and finite it really is. We have begun to appreciate the awesome interconnectedness of the sea world that makes all marine life everywhere subject to our actions. We have learned, for example, that DDT used in Africa has appeared in the tissue of Antarctic penguins thousands of kilometers away.

Moreover, we know now that we have vastly overestimated the sea's ability to feed our endlessly expanding human population. Although the sea is rich with life, by far the greatest part of it is not nearly as productive as the land. The open sea produces only about 100 grams of dry organic matter per square meter per year. Rich, near-shore upwelling areas may produce 2,000–4,000 grams per year in their surface waters. Only the coral reefs, scattered and difficult to exploit, reach levels of up to 4,900 grams per square meter per year.

By contrast, cultivated land in the temperate zone produces 4,000 grams per square meter per year, and in the tropics up to nearly 10,000 grams per year. On the average, a given land area will outproduce a comparable ocean-surface area by a ratio of nearly ten to one. Indeed, if all the assimilating algae (tiny plants which form the basis of all oceanic food chains and marine oxygen generation) were spread out on the surface of the ocean, the layer would be but two or three millimeters thick.

Despite its immense volume, much of the sea is comparatively empty, devoid of most life besides the thinly scattered showers of plankton falling from the surface zones. The overwhelming bulk of life in the sea is concentrated in a sensitive and narrow band along the margin of the continents, in coral reef zones, and in areas where abyssal oceanic currents rise as they meet continental slopes.

Imagine an island society living off the sea. If the surrounding seas were filled with fish, the islanders' future would seem quite bountiful. If, however, there were fish life only in a narrow, 50-meter band around their island, the future might suddenly seem precarious indeed. This is a realistic picture of our relationship to the oceanic life off our coasts.

These pages will concentrate on those underwater wildernesses that surround the islands and line the coasts of the world's tropics. There are several reasons for this perspective: the sublime beauty of these regions, their relative accessibility to academic study, and their availability to laymen through snorkeling and scuba-diving.

There is of course a large body of scientific material on this subject. This book, however, is directed primarily at the general reader. Its aim is to create a work of interest and value to expert and layman alike. Beyond that, the hope is that this illustrated record will arouse such an interest in the underwater wilderness that the reader will join those who are trying to defend it.

Part One

The Society of the Underwater Wilderness

The Forces that Shape the Underwater Wilderness

An infinitely complex interaction between two forces affects all natural life in the sea. We might characterize this interaction as the drive of biological forces toward essential order on one hand, and of geological forces toward endless change on the other. This is of course an oversimplification and a different approach from that which many scientists might take. It is not meant to supersede more technical explanations but to create a rational framework with which to visualize the basic processes of the sea.

The driving force toward stability and biological order is enormous, representing eons of survival by hundreds of thousands of species. For most species there is clear survival advantage in environmental stability. Each species then has a chance to evolve in ways which fit it more efficiently into the life of its community.

At the same time, geological forces place all life in the sea under constant pressure to adapt, to cope with change. These geological forces operate on two distinct time scales. One of these is within the realm of everyday experience: oceanic currents, tides, hurricanes, monsoons, typhoons, and waves.

The other of these operates far beyond ordinary human experience and is on what we might call earth-time, with thousands or millions of years between significant events. These long-term forces are described in the theory of continental drift and plate tectonics. This theory, first proposed in the 1920s by the German scientist Alfred Wegener, postulates that an original supercontinent, Gondwanaland, fractured into continent-sized plates which, over a period of 200 million years, have floated on a sea of dense molten lava to their modern positions. The motive force behind the movement of the plates is the continual welling of fresh lava from the earth's molten interior through fracture zones between the plates.

Obviously this movement is not directly observable. However, the colossal energies unleashed as the edges of these continental plates encounter each other produce such planetary spectaculars as earthquakes, volcanic eruptions, and the raising of mountain ranges. As we shall see in our chapter on corals, some of these planetary events were so cataclysmic that they snuffed out significant portions of all the life in the sea.

While we have studied and even observed certain effects of these earth-time forces, a lifetime is simply too short for us to experience many events, even minor geological ones. Our history records catastrophes such as the explosions of Thera in the Mediterranean, and Krakatoa in Indonesia. Recently, the world has been subjected to stupendous earthquakes in Anchorage, Alaska and in Guatemala. In the latter an estimated 23,000 persons died and 77,000 were injured. This disaster illustrates all too well the immensity of continental forces, for in earthwide terms the Guatemala quake was only a two-meter movement between the American and Caribbean plates. It

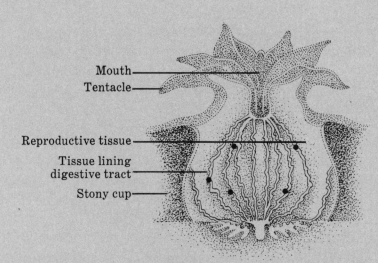

Cross section of a coral polyp, the living tissue of the coral.

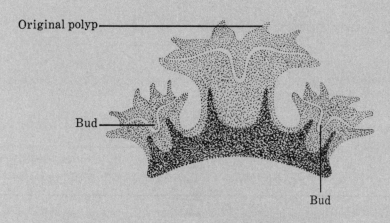

Formation of coral polyps through budding.

was a minor relief of tectonic stress—an infinitesimal occurrence—compared to such past events as the raising of the Himalayas.

The interaction between biological and geological forces colors every activity we observe in the underwater wilderness. Coral colonies provide an excellent illustration of the range of phenomena these interactions produce. In a process of exacting, unending growth uncounted billions of individual coral polyps are born and live out their lives in colonial limestone tubes. They lengthen each tube slightly by secreting limestone they extract from sea water. Then they die, leaving the tubes to be further lengthened by later occupants of the colony. The result of this tiny drama, repeated an astronomical number of times, is the wonder of lush coral reefs.

And yet, even leaving the effect of man aside, all coral reefs suffer constant damage. Like all biological entities, they must adapt to a range of conditions, always facing inherent limits in their ability to survive. The shallow tops of the coral heads may be exposed to air during extreme low tides, and coral polyps suffer fatal desiccation when exposed to air for any extended period. They are affected by a relentless succession of waves, which in a storm can overturn and destroy entire colonies. They can be suffocated by silt carried into the sea by rivers after monsoon rains. They can be destroyed by earthquakes or volcanic eruptions. Particularly in the Pacific, their underlying land mass may simply sink into the sea far beyond the reach of life-giving sunlight. It has now been accepted that important coral structures of today known as barrier reefs, atolls, and guyots can only have resulted from this "subsidence," due to tilting of their supporting continental plates.

Such destructive occurrences, however, illustrate another important principle. While local extinctions occur, biological and geological forces over a long period tend in general to assist the survival of species. We will see why this occurs as we consider another of these forces.

For instance, when we enter shallow coastal waters anywhere in the tropics, we can readily identify familiar inhabitants. The interplay between biological history and such phenomena as oceanic currents makes inevitable the fact that we find groupers, blennies, triggerfish, and other ecologically related fish around island after island in the tropical world. This world distribution of species is one of those designs in nature remarkable for its simplicity and success. The gravitational effects of the earth, sun, and moon set the waters of the earth in constant motion. One result is a network of oceanic streams in which masses of water move regularly about the earth in a fairly well-understood system of currents. These compensating flows of water affect the world's climate in major ways. A classic example is the Gulf Stream, whose warm waters make western Europe's climate far milder than it would otherwise be.

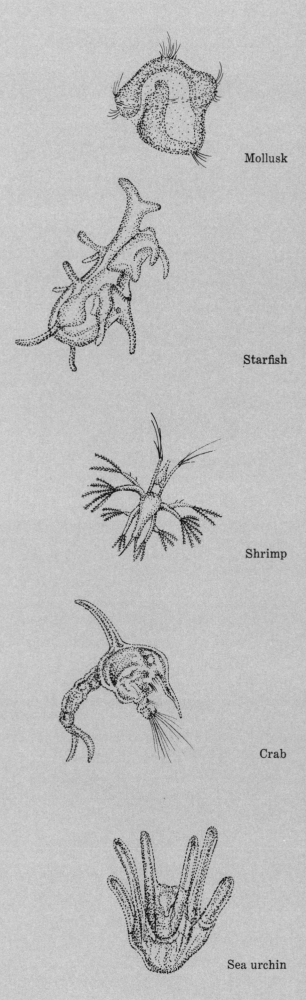

Mollusk

Starfish

Shrimp

Crab

Sea urchin

Larvae. In their pelagic or free-floating period, these marine organisms are transported to new homes.

Given this planetwide movement of water, only one other phenomenon is required to successfully seed the planet with marine life: the members of a species have to pass through a pelagic or free-floating period during which they can be transported to new homes. This is precisely what we do find in many widely differing forms of marine life, including crinoids, groupers, and eels. The pelagic period is most frequently an animal's larval or juvenile stage, at the end of which it settles in its new home. The only limitations on how far a species may spread is the length of this pelagic interval and whether the individuals can survive the journey. Thus, tropical Pacific species do not spread into the Atlantic due to the cold-water barriers of Cape Horn and the Cape of Good Hope. This continuous movement of individuals in a species has two profoundly stabilizing consequences for the species as a whole. First, if a local catastrophe such as a volcanic eruption or an earthquake snuffs out the species in one locale, the ever-moving currents will bring fresh stock. Second, new islands or substrates formed from the same causes will begin to be stocked almost immediately. Other important effects are the continual introduction of fresh, new generations worldwide and the carrying away of juveniles so they do not compete or inbreed with their parents. Though it is true that seeding will assure that groupers, corals, and others will be brought to new sites, there is no guarantee they will forever resemble their forebears. They will evolve to suit the new conditions they find. Thus members of the same genus often differ from one major area to another. For example, the royal gramma (*Gramma loreto*) is a small, brilliantly-garbed relative of the larger groupers and sea basses (Illustrations 7–10). This small fish hovers above the protective coral, snatching bits of nourishment from the drifting food stream. When threatened, it darts into a coral hole or crevice. In the southern Caribbean the purple coloration of the gramma's body extends for fully half its length (Illustration 7), while in the northern Caribbean the bright yellow tail colors extend nearly to the fish's head (Illustration 9). These local color variations have evolved since the Isthmus of Panama rose more than a million years ago, severely reducing the species-spreading currents in the Caribbean. The variations are a perfect example of how members of the same species evolve local changes through isolation.

A far broader example is found among the butterfly-fishes (Illustrations 11–27). These Chaetodontidae, whose name derives from the Greek words for "hair tooth," are among the most conspicuously colored members of the coral reef community. Their lavishly hued, disc-shaped bodies ablaze with colors and stripes, they are usually the first fish we perceive in any coral waterscape. As one moves from one major reef area of the world to another, one finds old friends and local strangers among these butterfly-fishes. The old friends are the species whose free-floating larval stage is able to survive longer journeys to new homes. The local strangers, or endemic species, have evolved in isolation without continual restocking from the parent species.

The epic interaction between biological and geological forces has continued since the dawn of life on our planet, and barring destructive actions by man it should go on for further eons. We must remember that the world as we experience it in a lifetime is but an instant in earth history. Europe and America move a few centimeters apart in a lifetime. An island called Surtsey rises in flame and thunder from the north Atlantic. Guatemala is devastated by an earthquake. Somewhere on earth the last members of a species are extinguished. But earth abides, and I am never so cognizant of history's span as when a geologist tells me that not so long ago in earth-time, the present site of Harrisburg, Pennsylvania was 10,000 meters underground.

We do have the advantage, however, of looking backward over earth's history with our growing science and technology. We study biology, geology, astronomy, paleontology, physics, and chemistry, and bit by stubborn bit our historical antecedents emerge from the past. Look at the stir the coelacanth caused when it leaped out of prehistory into a fisherman's net. Who knows how many more surprises we must face before we can accurately say what happened through out past ages?

Our earth history is like some planetary textbook whose chapters undergo never-ending expansion and change. Our successors a thousand years hence will still be coping with gaps in their knowledge and with the embarrassment of substituting new explanations in place of long-accepted ones found wanting. It is that exhilarating process of discovery that drives us ever to seek new insights. Man's curiosity itself must be classified with the great driving forces of the world. If our understanding can only keep ahead of our economic and population pressures, we may someday live in harmony and balance with the incalculable forces that shape us.

6. *A fallen palm tree, mute reminder of the power of wind and storm.* (*Huahine, French Polynesia*)

7. *The southern Caribbean variant of the royal gramma* (Gramma loreto). *This specimen bears its purple coloration over more than half its body length.* (*Netherlands Antilles, Caribbean*)

8. *Portrait of the royal gramma.* (*Curaçao, Caribbean*)

9. *The northern Caribbean color variant of the royal gramma, whose purple coloration extends less than halfway along its body.* (*Cayman Islands, Caribbean*)

10. *A group of royal gramma hovering above their sponge-encrusted coral refuge.* (*Cayman Islands, Caribbean*)

11–27. *A selection of butterfly fish* (Chaetodontidae) *from around the world.*

11. *The long-snouted butterflyfish* (Forcipiger flavissimus), *discovered by Captain Cook.* (*Hawaii*)

12. Chaetodon ephippium, *a saddleback butterflyfish.* (*Australia*)

13. Chaetodon trifascialis, *a shield-shaped butterflyfish.* (*Rangiroa, French Polynesia*)

14. Chaetodon ornatissimus, *an ornated butterflyfish.* (*Kona, Hawaii*)

15. Chaetodon quadrimaculatus, *a brightly-colored butterflyfish.* (*Rangiroa, French Polynesia*)

16. *Reticulated butterflyfish* (Chaetodon reticulatus). (*French Polynesia*)

17. *Four-eyed butterflyfish* (Chaetodon capistratus). (*Caribbean*)

18. *The red-tailed butterflyfish* (Chaetodon chrysurus), *found in profusion on reefs of the Red Sea.*

19. *The masked butterflyfish* (Chaetodon lunula), *widely distributed through the Indo-Pacific.* (*Australia*)

20. *A banded butterflyfish* (Chaetodon striatus). (*Caribbean*)

21. Chaetodon sp., *discovered in the Galápagos. This species is new to science.*

22. *A spotfin butterflyfish* (Chaetodon ocellatus). (*Caribbean*)

23. *The saddled butterflyfish* (Chaetodon falcula). (*Fiji Islands, South Pacific*)

24. Chaetodon auriga, *a threadfin butterflyfish.* (*Truk, Micronesia*)

25. Chelmon rostratus, the beaked *butterflyfish.* (*Great Barrier Reef Province, Australia*)

26. Chaetodon pelewensis, *a dot and dash butterflyfish.* (*Hawaii*)

27. Heniochus monoceros, *an extremely shy species of butterflyfish, very difficult to photograph successfully.* (*Red Sea*)

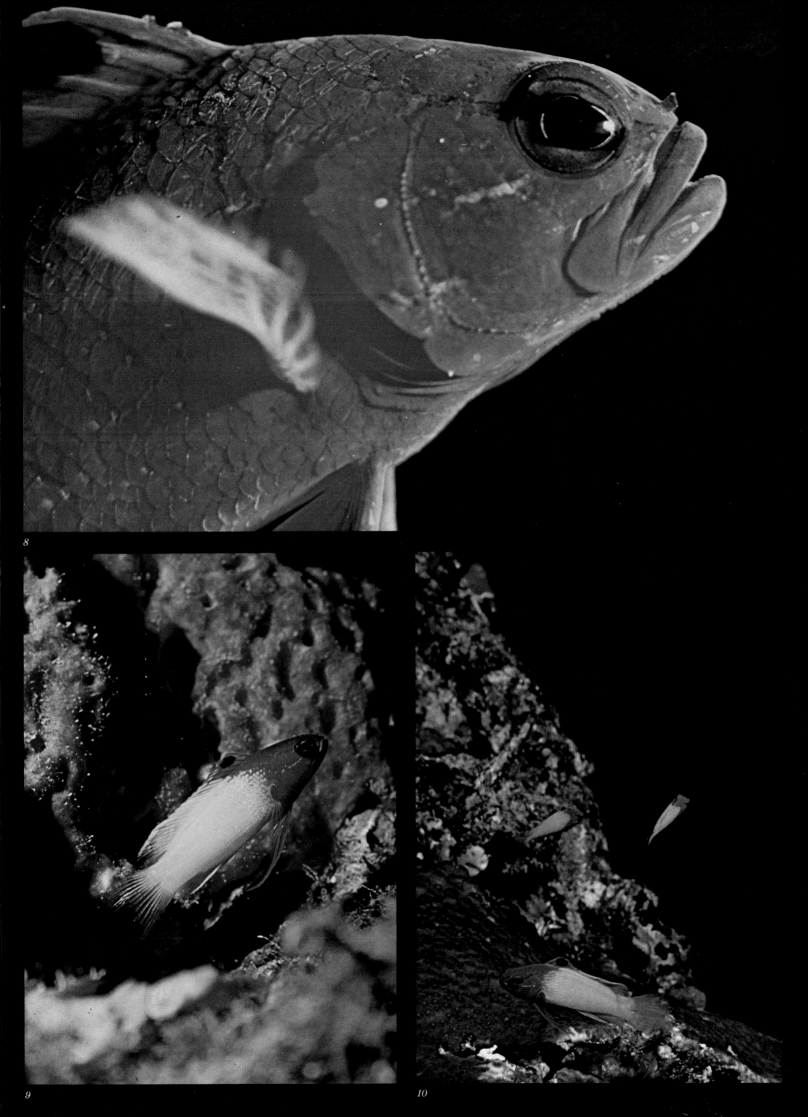

8

9   10

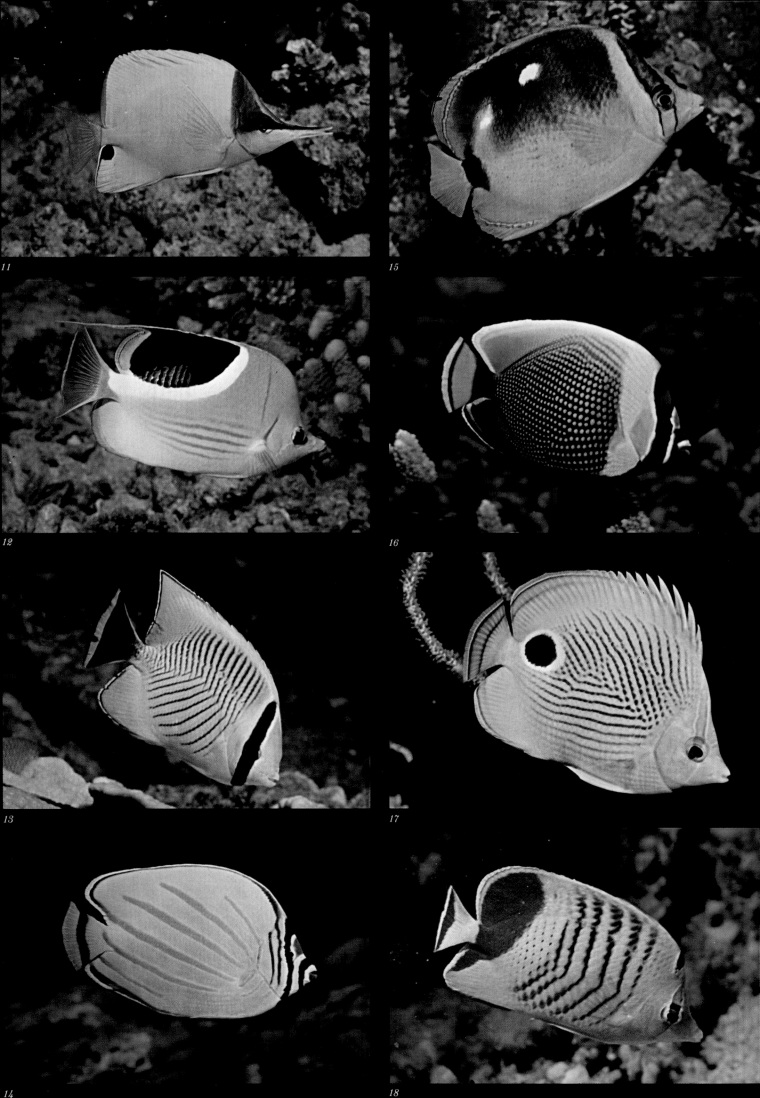

11

12

13

14

15

16

17

18

19

23

20

24

21

25

22

26

Corals, Reefs, Islands, and Atolls

Coral in its many guises is an endless source of wonder and delight. Since it plays a critical role in the underwater wilderness, we should consider how it does so. At the basic level is the coral polyp itself, a complex and fascinating animal, some of whose functions are still not completely understood. At another level are the structures within which the individual polyps grow as they form a colony, and the distribution of those colonies. In addition, there are structures formed by large aggregations of coral colonies: reefs, islands, and atolls.

We should also study corals in terms of the part their prehistoric ancestors played in ancient seas as well as investigate them in terms of their present role as integral members of the complex modern reef community.

The "gardens" of coral in shallow tropical seas are not, as they may appear, plants. This discovery was made in 1753 by J. A. de Peysonell, who realized after thorough study that the coral structures were built by aggregations of minuscule tentacled animals. These coral polyps have cylindrical bodies with discs at the upper and lower ends, mouth openings in their upper discs leading to digestive tracts, and a circlet of tentacles surrounding the mouths to capture small particles of food from the surrounding water (Illustrations 30, 133). These tentacles are armed with contact-actuated stinging cells called nematocysts (Illustration 130).

This method of capture is common to thousands of tentacled organisms in the sea. If there were no more to the coral polyp than this, it would merely be another beautiful reef-citizen struggling, with countless competitors, for the largesse of the food stream. But the coral polyp is unique in other ways. Its outer skin is able to extract soluble calcium ions from the surrounding sea water. It then secretes a calcium carbonate skeleton on the outer surface of its basal disc and lower body, much as the body of an oyster secretes a protective shell. Since the polyp has a cylindrical body, the resulting limestone structure is a cup or tube into which the coral polyp can withdraw for protection.

The rate at which the polyp deposits calcium carbonate varies in different species and is highly dependent on sunlight. Corals are able to produce daily some 10 grams of limestone per square meter of polyp surface. This conversion process consumes so much energy that, when measured by oxygen consumption, a coral polyp requires more than twice the energy a man does per gram of body weight. Clearly, the coral polyp uses these high levels of metabolic energy in the calcium conversion process. For a long time it was assumed that this energy came from the metabolic conversion of plankton upon which the polyps fed. The fact that coral polyps open to feed at night when plankton is most plentiful seemed to support that assumption. In recent years, however, scientists have discovered that coral does not thrive solely by feeding on plankton. Indeed, certain corals can be totally deprived of plankton and continue to grow. We now know that this is possible because imbedded in the tissue of these coral species are single-celled algal symbionts known as *zooxanthellae*.

These algae manufacture oxygen and carbohydrates through photosynthesis, furnishing the polyp the oxygen for its limestone secretion process. At the same time the algae break down the polyp's expelled carbon dioxide and nitrogenous wastes. When maintained in darkened aquaria, coral species associated with the algae cease growing. Those whose metabolisms seem to function on captured plankton continue growing in darkness, but less so than when illuminated.

Many questions remain: Why would corals retain both plankton-trapping body mechanisms and symbiotic algae? Why do reef corals thrive in the plankton and nutrient-filled turbulent outer waters of their reef structure and cease developing on the lagoon (interior) side? The answers undoubtedly involve both of these feeding methods. Perhaps in some species they perform complementary functions and in others operate as back-up systems to assure the coral's survival.

Coral Colonies   The building of coral structures by colonies of polyps is a natural product of the polyp's reproductive system. Corals reproduce both sexually and asexually, that is, both through the union of released spermatozoa with eggs and through budding. Each process plays an important part in the development of coral structures and their dissemination to new locations. In sexual reproduction the eggs and sperm during the breeding season develop in swollen tissue along internal skin folds known as mesenteries. When the spermatozoa are ripe, they are released and drift until they are drawn into the mouth of another polyp some distance away. Within this second polyp the spermatozoa fertilize the ripe eggs. The fertilized eggs are soon released as tiny, flat-bodied larvae with microscopic surface hairs whose rapid beating results in a limited mobility. These larvae (called planula) may drift from one to several weeks, carried by oceanic currents for considerable distances. This pelagic period accounts for the spread of coral reefs and the reseeding of damaged reef areas. Eventually, some of the drifting larvae come in contact with sufficiently solid substrate. They then attach themselves by a cement-like substance and begin adopting a tubular shape, developing tentacles around the mouth and secreting their protective limestone cup.

While sexual breeding helps distribute corals, asexual breeding, or budding, is the basis on which coral colonies and structures develop. In this process the first coral polyp develops a swelling "bud" or daughter polyp which soon secretes its own coral cup. Eventually these two polyps form more, resulting in a cluster, or colony, of polyps. The pattern in which

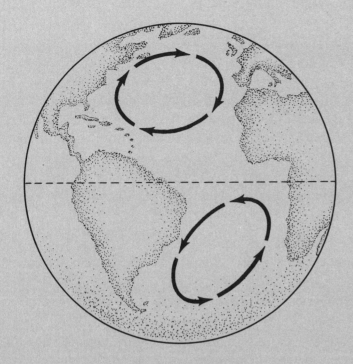

Ocean currents are affected by the sun's heat, winds, and the Coriolis parameter (also called the Coriolis force). This force, a result of the earth's rotation, deflects ocean currents to the left in the southern hemisphere and to the right in the northern hemisphere.

each species buds and builds accounts for such widely divergent colony shapes as the domed, ridged brain coral (*Lobophyllia*), on one hand, and the delicately branched staghorn coral (*Acropora cervicornis*), on the other (Illustration 41). Other coral relatives, such as gorgonia, use other body-constructing mechanisms.

Corals are widespread across the oceans of the earth. They are not restricted to the tropics or to shallow water. I have seen solitary corals dredged from a depth of over 300 meters in the Caribbean, and there are accounts of colonial corals occurring in both deep and cold waters.

However, the true reef-corals are limited both to a narrow range of preferred, stable temperatures and to clear water less than 50 meters deep. These corals, as we have seen, need the penetration of strong sunlight for the photosynthetic processes of their symbiotic zooxanthellae. Since the sun's penetration is drastically reduced at a depth of 50 meters even in fairly clear tropical waters, the activity of the algae —and hence coral growth—diminishes sharply at those depths.

While the reef-building corals are generally found within the limits of the sun's winter and summer elevations (23.5 degrees north or south latitude), they are not found on all coasts within that span. This is due to another powerful influence: the great oceanic currents, plying their endless routes.

The ocean currents influenced by the earth's rotation generally move westward at the equator and are heated by the tropical sun. Encountering land masses, the currents turn northward above the equator and southward below it due to the Coriolis force, the frictional effect of the earth's rotation. Thus, warm waters bathe the eastern shores of the continents and their associated offshore shallows. These areas account for a large portion of the major coral reef concentrations of the sea.

Extending our oceanic current analysis, we find that near the poles, the major currents turn eastward and become cooler. When they once again encounter the barrier of land masses, they turn southward above the equator and northward below it. This causes extremely cold polar waters to bathe the western continental slopes. When these equator-directed currents reach the continents, they not only change direction but also, in localized areas, raise great masses of deep water and accompanying nutrients (products of decomposed tissue falling into the depths) into the shallows. This upwelling greatly increases the concentrations of life-supporting plankton in the continental shallows and results in enormous populations of fish life off such coasts as those of Chile and Peru. However, these waters are too cold for widespread coral development. These ocean-wide forces thus limit reef coral growth to the eastern shores of continents and the islands of the open tropical seas, the warm-water zones.

Species of coral differ in their rates of growth. Some of the very dense dome corals such as *Porites* may increase their weight by 100 to 200 percent annually, while the growth rate of an intricate branched thicket of *Acropora* may reach 400 percent.

Thus, those corals which metabolize most efficiently, thereby secreting the greatest quantity of limestone, gradually dominate the active growth zone of the reef. These tend to be corals which have evolved most recently and become most specialized. This specialization involves some degree of risk, however, for if ecological conditions change, it is the older, slower-metabolizing corals which best survive. The newer corals, best adapted to prevailing conditions, may be the first to perish if those conditions change.

Kinds of Reefs

Coral reefs tend to originate in shallow waters just off a land mass. Here coral larvae, carried by currents, find solid substrate, attach themselves, and begin to form colonies. This results in the formation of a narrow ribbon of growing coral running parallel to the coastline. The restless currents bring fresh coral larvae and other animals that, in concert with the corals, constitute a reef fauna: algae, sponges, mollusks, reef fish, and others. Meanwhile, the surf and sun provide oxygen, nutrients, and sunlight to enable the coral colonies to flourish (Illustration 44). The ribbon of coral thus spreads along the coast and becomes broader as the coral colonies grow larger and more profuse.

The most intensive coral growth occurs on the seaward side of the colonies, for it is here that the high-metabolism, efficient corals find the necessary conditions for growth. On the leeward or coastal side, meanwhile, there is a gradual dying off of all but the hardy, slow-growing species. The water in the lagoon between the coral reef ridge and shore is still. Corals here are now protected from the vital surf by a barrier consisting of the limestone bodies of their relatives. Over time, the original lagoon colonists have their limestone structures ground down to fine calcareous sand and debris by such limestone processors as parrotfish. These parrotfish with their massive, beaklike jaws eat the coral, passing limestone as beach sand in prodigious quantities (Illustration 58). Researchers have estimated that a single medium-sized parrotfish passes more than a ton of sand in a year. Little wonder that great strands of sandy beach and sandy shallows are found in the vicinity of lagoons.

The reef we have been describing is known as a fringing reef in that it usually follows close to a coastline. Young reefs tend to be of the fringing type. More mature reefs, as we shall see, frequently become barrier reefs, running parallel to their shore but at a considerable distance from it. The lagoons or channels thus created may vary in width from a few hundred meters to 150 kilometers and reach depths of 60 meters.

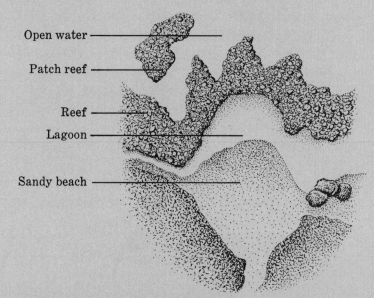

Open water

Patch reef

Reef

Lagoon

Sandy beach

A coral reef shown in relation to its landward side and to its seaward side.

Aerial view of the development of a coral atoll. Although an atoll is on the surface a circular reef around a lagoon, it actually covers a submerged volcanic island.

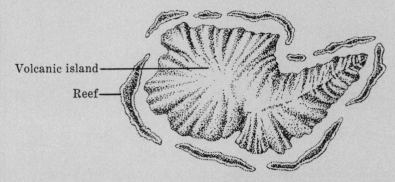

Geologically young volcanic island with a fringing reef

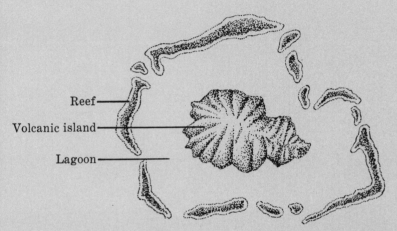

Mature volcanic island with a barrier reef

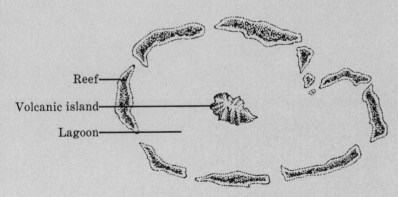

An even older volcanic island, almost fully submerged, surrounded by a reef forming what is known as an "almost atoll"

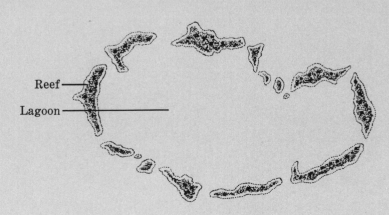

A completely developed coral atoll

Barrier reefs are extremely complex. The effects of storms, weathering, and fish processing of coral limestone combine to build sandy strands which may in time break the surface. Over a long period these are seeded by terrestrial flora such as coconuts and mangrove and achieve permanence. Many of these islands in turn develop their own seaward fringing reefs. A third classic reef structure is the atoll.

These are in many ways the most interesting, occurring as they do far from continental mainlands and surrounded by water that may be thousands of meters deep. Atolls are roughly circular, with steep outer sides and a fairly shallow lagoon. They also have passes or cuts in the ring structure, through which great volumes of tidal water flow.

For many years it was assumed that atolls simply grew upward from very deep water by the accumulation of countless generations of polyps. It was not until Charles Darwin's publication of *Structure and Distribution of Coral Reefs* (1842) that a comprehensive theory on coral reefs was proposed. Darwin's research showed that the growth of reef corals is limited to waters less than 50 meters deep. He coupled this fact with subsidence, the sinking of entire volcanoes or coasts which we now know is due to tilting of their supporting continental plates. This process occurs so slowly that sometimes reefs manage to keep growing and become the structure that remains even after the volcano or land mass has disappeared. This combination of rapid shallow coral growth and subsidence explained both atolls and barrier reefs. Simply, Darwin postulated an initial fringing reef stage about a volcanic or coastal mainland. As the land slowly subsided, coral grew outward and upward. If the subsidence were not too rapid, the coral growth could keep pace, the result eventually being a deep, wide channel and a barrier reef. If the land were a volcano, it would eventually disappear beneath the surface, and the circular barrier reef would now be an atoll.

Darwin's theory that growing corals paced subsiding land masses received a setback when explorations by *H. M. S. Challenger* in 1872 to 1876 revealed submarine mountains that did not bear coral. Further expeditions made borings whose results were interpreted in various ways by partisans of different theories.

One of the most important theories, advanced in 1919 by Reginald A. Daly, was that of glacial control. He proposed that during the last ice age the continents were covered by gigantic ice caps containing such prodigious amounts of water that the sea level was lowered by some 55 meters. This killed all previously formed coral reefs within that range. The former ocean bottom, exposed by the withdrawal of the water, was now subjected to wave action. The waves cut through the exposed limestone, forming broad platforms off the coasts of high land masses. This process can be observed today on such Caribbean is-

Cross section of the development of a coral atoll.

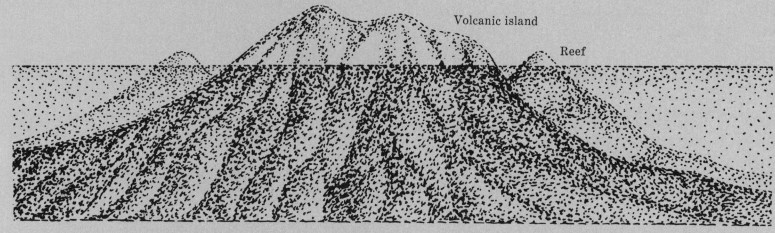

Volcanic island

Reef

Young volcanic island with a fringing reef

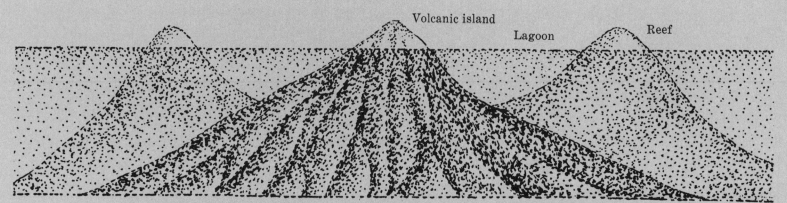

Volcanic island

Lagoon

Reef

Mature volcanic island with barrier reef

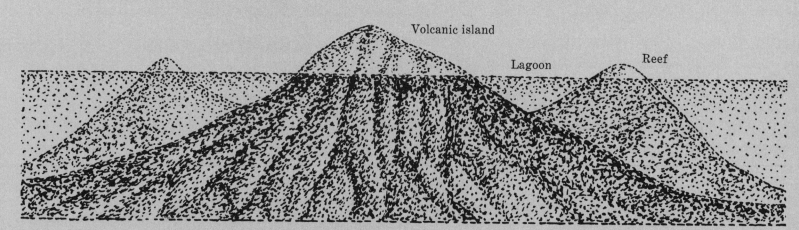

Volcanic island

Lagoon

Reef

An even older volcanic island surrounded by a reef forming an "almost atoll"

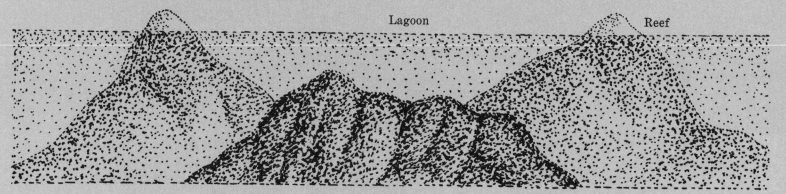

Lagoon

Reef

Completely developed coral atoll

lands as Bonaire. Here former sea levels are marked by shelves cut into the coral rock of the island mass. Eventually the undercut limestone cliffs collapse and are broken up, leaving a flat platform just beneath the surface. On Bonaire at least three former sea levels are visible above today's level, with another one at a depth of about 50 meters beneath the surface of today's Caribbean.

Daly's theory next postulated that as the sea levels rose at the end of the ice age, new corals grew on the inundated platform. By Daly's argument, all modern coral reefs are less than 10,000 years old. This theory accounts for many cases of reef structures which had seemed to be exceptions to Darwin's theory, particularly in rising sea floor areas such as the Bahamas. New theories and discoveries have followed in the wake of Darwin's and Daly's interpretations. During World War II, an American submarine commander, H. H. Hess, discovered nearly 160 underwater mountains whose tops ranged from 1,000 to 2,000 meters below present sea level. Hess called these mountains guyots, and suggested that their flattened tops were the result of wave action. He further suggested that they might have been drowned by sea levels rising through the accumulation of sediments.

In 1951–1952, in anticipation of atomic tests at Enewetak atoll, seismic tests were undertaken to assess the effect of the weapon. These borings penetrated to a basaltic foundation after passing through more than 1,200 meters of coralline limestone. During the 1950s, scientists also made a major discovery: the Mid-Atlantic Ridge and other oceanic ridges around the world were the sites of the geologically youngest rock on the sea floor. Moreover, the farther from these seismic ridges that sample rocks were collected, the older they proved.

These discoveries resulted in the acceptance of the theory of continental drift. In this logical construct, the mid-ocean ridges are the site of upwelling molten rock, which hardens to become new crustal material. This is added to the continental plates, upon which our continents and ocean basins slowly move. These new geotectonic theories have also given rise to four new classifications of coral reefs. Atolls are thus logically identified as oceanic volcanic reefs. The reefs of active volcanic zones are now classified as mobile belt reefs. The reefs of the Bahamas are quasicratonic—a result of rising continental plates—while the Great Barrier Reef is an epicontinental reef. The vindication of Darwin's theory, which had long been open to question because there was no adequate explanation of the subsidence he postulated, has taken over one hundred years of controversy and research.

Coral History   As a further insight into coral, we might briefly examine its antecedents and its historic survival despite severely changing earth conditions. It is, for example, now known that at least four times during

the past two billion years, major extinctions of much
of the life on earth occurred. At each juncture, dominant species emerged.

Two billion years ago, there began large-scale reef-
building by blue-green algae in layered structures
now called stromatolites. These reefs were associated
with primitive mobile sea creatures called archaeocyathids, and this simple but stable reef community
spread widely over the earth. Then, at the end of the
Cambrian Period (600 million years ago), all
archaeocyathids were extinguished by some unknown
global event.

In the Ordovician Period (500 million years ago)
new reef-dwellers emerged, and with them a more
complex reef society. Calcareous sponges, bryozoans,
and other forms burgeoned with the colonial algae;
and, most important for our purposes, two orders of
coral appeared. Reef builders flourished. During the
Silurian and Devonian Periods (430–395 million
years ago), enormous reef communities filled the
shallow seas. The Age of Fishes flowered during the
Devonian, which was then followed by another universal mass extinction, with very little reef life
surviving.

The following Carboniferous and Permian Periods
(345 and 280 million years ago, respectively) saw an
entirely new undersea community evolve, with dominant crinoids (feather starfish), bryozoans, and
others. Once again calamity struck, with the greatest
cataclysm ever to affect marine life. More than 50
percent of all existing families died out.

And yet a new period of recovery began, the Mesozoic
or Age of Reptiles (225 million years ago). Then at
the end of the Mesozoic still another cataclysm
occurred. This time about one quarter of all families,
including all the dinosaurs, were extinguished. These
cataclysms, the source of much speculation, may not
all have resulted from the same cause. The catastrophe that closed the Paleozoic, for example, may have
been a worldwide explosion of volcanic activity.

We have thus far considered corals as functional
organisms and as builders of exquisite self-protective
structures. As a group, we understand their role in
building living breakwaters that protect continental
coasts, and massive citadels such as atolls. In subsequent pages we shall see coral in two other roles: as
hunter and prey in the endlessly active ocean society;
and as home, shelter, and community for a myriad of
species seeking protection. We shall see the reef as
the stage for the dramas of other reef-dwellers.

28. Coral colonies from Australia's Great Barrier Reef. They nearly break the surface, thus forming a vast maze of hazardous barriers throughout the area.
29. An aerial view of Bora Bora, regarded by many as the world's most beautiful island. Its large interior lagoon has only one pass to the sea and is thus protected from the sea waves. (French Polynesia)
30. A polyp of the Caribbean coral (Eusmilia fastigiata) at sunset, as its tentacles have begun to emerge. It measures about two centimeters across the widest part of its body.
31. The Caribbean gorgonian known as the sea whip. It occurs in great numbers in the shallow waters off the islands.
32. Gorgonians, relatives of coral. They create a flexible rather than a stony skeleton. (Galápagos Islands)
33. A large gorgonian, Melithia squamata. (Great Barrier Reef, Australia)

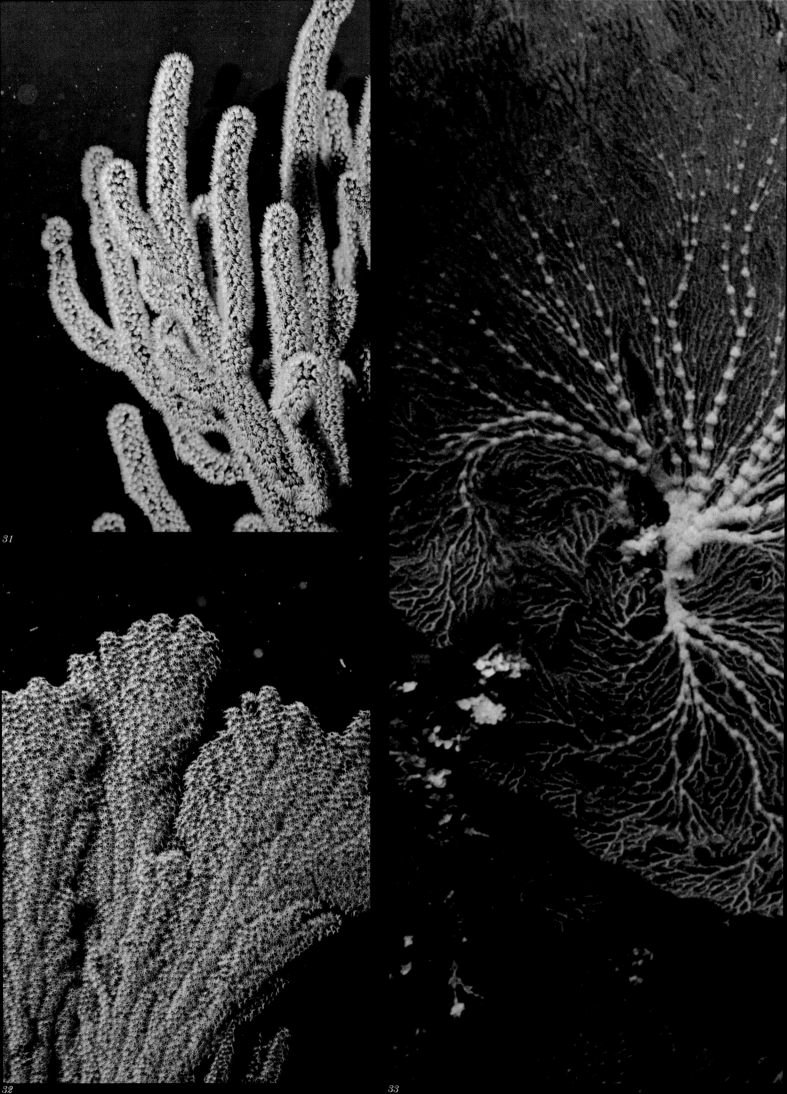

31

32

33

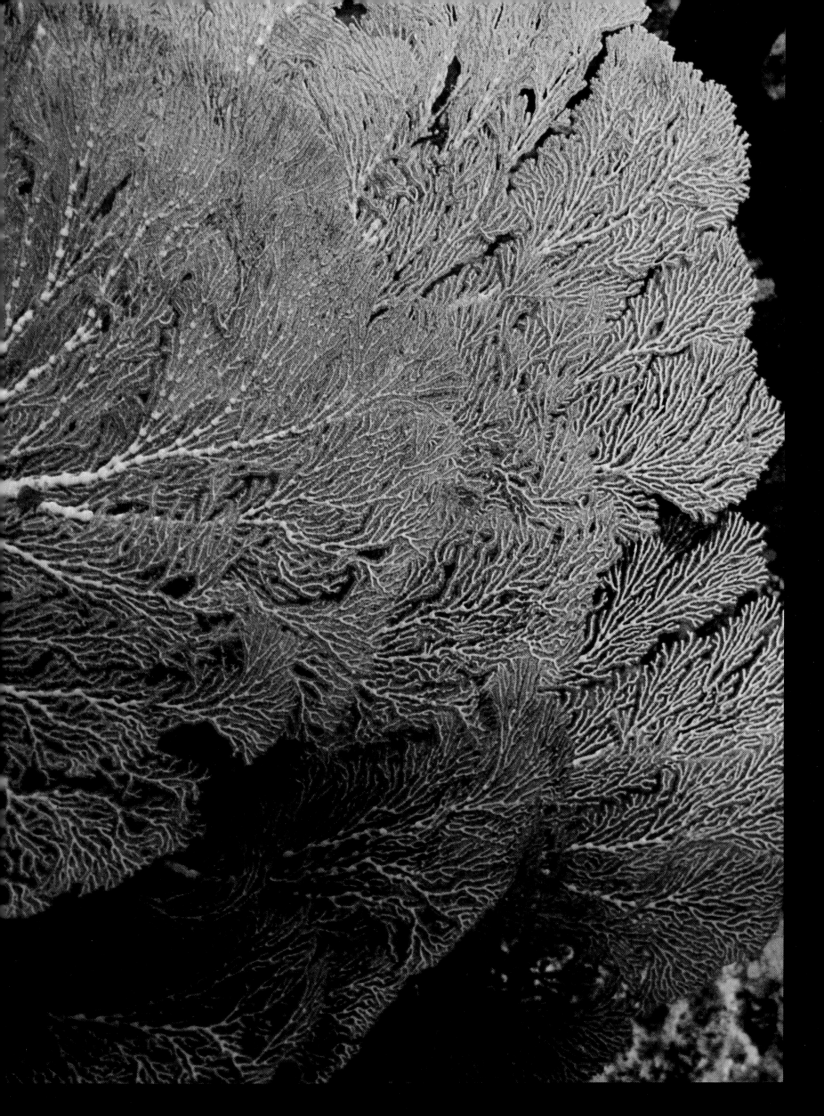

34. Agaricia agaricites, *a Caribbean coral that builds vertical plates several centimeters in height.* (*Bonaire*)

35. Melithia squamata *at night, sheltering a small damselfish and wrasses.* (*Great Barrier Reef, Australia*)

36. *The beret coral* (Scolymia)*, a solitary coral found at depths of 15 to 50 meters.* (*Curaçao, Caribbean*)

37. *A gorgonian sea fan. They are especially abundant on the shallow reefs of the Cayman Islands.* (*Caribbean*)

38. *Branched processes of a colorful coral colony found in shallow waters of the Great Barrier Reef Province.* (*Gannet Cay, Australia*)

39. *The fire coral* (Millepora alcicornis)*. It occurs in fingerlike clusters, as vertical plates, and as encrustations on other corals.* (*Cozumel, Mexico*)

40. Millepora alcicornis, *the fire coral, in the southern Caribbean.* (*Bonaire*)

41. Acropora cervicornis, *the staghorn coral of the Caribbean.* (*Cayman Islands*)

42. *The branching hydrocoral* (Stylaster)*.* (*Cayman Islands, Caribbean*)

43. *A wire coral with its polyps extended, photographed at night.* (*Roatán Island, Caribbean*)

44. *A variety of corals found on the reefs of Tahaa, French Polynesia.*

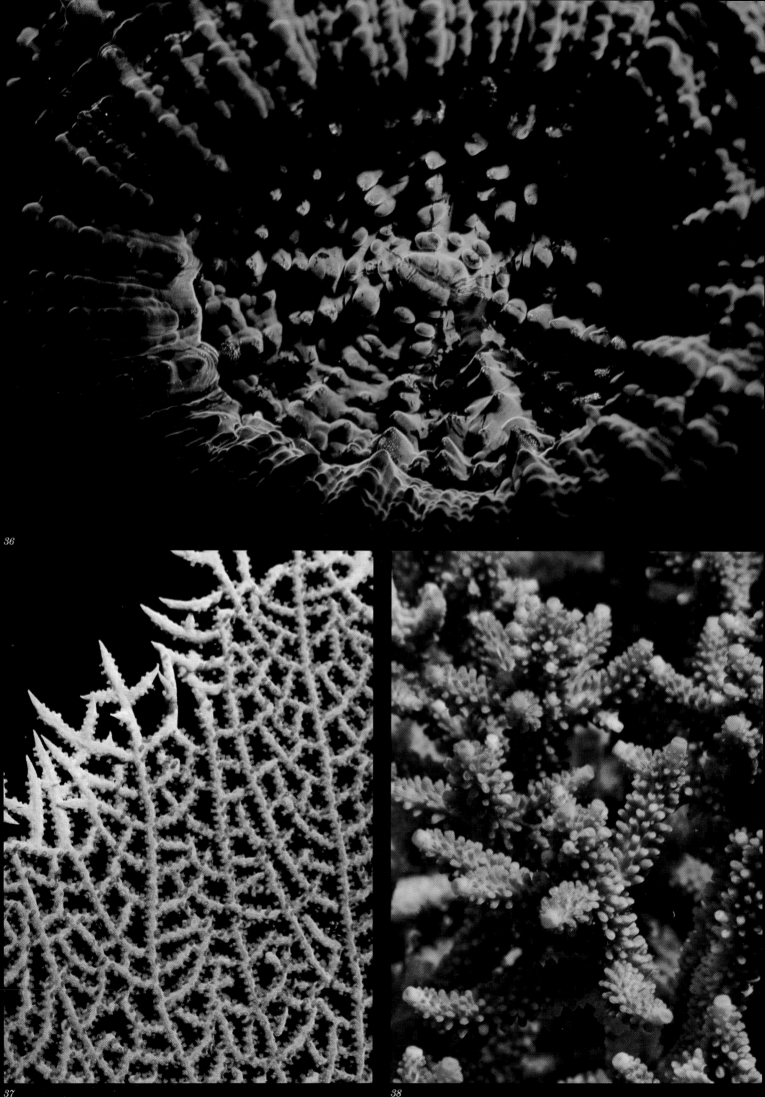

36

37

38

39

40

41

42

43

The Jungle in the Sea

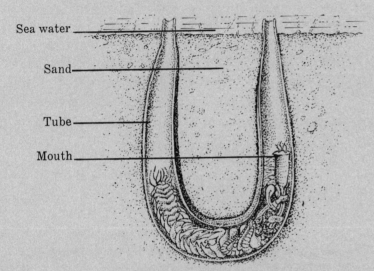

Sea water

Sand

Tube

Mouth

A Parchment Worm

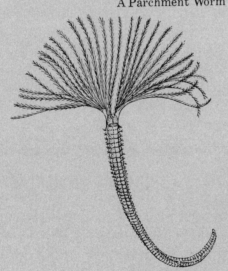

A Feather-duster Worm

Polychaete or tube worms, sea-going relatives of the familiar earthworm.

Above: A parchment worm, one of the serpulid type of polychaetes. It builds a parchment tube into which it retreats for shelter. Some serpulids range widely, sometimes attaching their coiled tubes to the shells of large mollusks. Below: A feather-duster worm, one of the spectacular sabellid type of polychaetes. These worms build leathery tubes as much as 50 centimeters long and extend sprays of feathery gill plumes for trapping food as well as detecting predators.

In the reef, as everywhere else in the sea, predation is the key to survival. It is always and everywhere a constant that determines the survival of each species and its continuing place in the community of reef dwellers.

The words "undersea predator" generally bring to mind one image—the shark (Illustrations 277–279). Yet in the intricate web of undersea society the shark is but one of many creatures whose lives are entwined in a ritual of kill or be killed, eat or be eaten. As our understanding of the complex reef society grows, we realize that the sea is a vast jungle, with hunter and hunted engaged in an endless round. From the smallest coral polyp to the largest of sharks, survival is the rule of life. For human beings, the fascination of this undersea jungle is the incredible range of methods and styles among its hunters. As we have seen, the countless coral polyps are all hunters, rooted throughout their lives to a single spot, stretching out their tentacles to capture plankton. Yet even as the coral polyps capture prey, they themselves are constantly being preyed upon: butterfly-fish, bristleworms, starfish, crabs, and a thousand other browsers consume the tiny reef-builders (Illustration 54). Corals are not the only sessile (attached) predators reaching into the food stream. Soft coral polyps, less toothsome than their stony-coral brothers, collect in water-inflatable, saclike colonies (Illustrations 253, 255) rather than lime-stone skeletons. Gorgonian polyps share a whiplike structure so flexible that they can survive the unceasing wave-surge of the shallows without breaking, or stand erect in the eerie quiet currents of deep slopes (Illustrations 33, 35, 43). One unusual family of sessile feeders are the polychaete worms (Illustrations 46–47, 50). The human visitor to the reef at first sees tube worms as a cluster of brightly colored feathers which instantaneously retract at a wave of the hand. Closer inspection reveals that there are two distinct types of worms. Serpulid worms commonly resemble a tiny, double-spiraled Christmas tree seemingly growing upon the surface of a large coral colony. Sabellid worms exhibit a single or bilobed rim of long feathery arms extending from a rather flabby stalk.

Examining these tube worms more closely, we find that the serpulid type secretes a hard limestone tube that in some species lies along the surface of the coral host but in others penetrates the colony. To protect their vulnerable bodies from predators, the sabellid worms secrete a mucus-and-sand tube. Yet, as predators themselves, all of these tube worms must extend their delicate arms to feed. Forever alert to the vibrations of approaching danger, the polychaete worms can retract their bodies in a flash and disappear into their tubes. Still, one will occasionally see a serpulid with the top of a feather-spiral nipped off.

Both families of polychaetes possess a segmented, wormlike body equipped with hairs, or cilia, to give the animal traction in its tube (Illustration 46). The upper body is surmounted with a mantle, or collar, which is involved both in reproduction and in the secretion of the protective tube. The feathers, or arms, are not only respiratory and food-gathering organs but also assist in sensing the vibrations of approaching predators. In serpulids such as *Spirobranchus giganteus*, the worm also possesses an operculum, a kind of trap door that seals the entrance to its refuge. Small horns or processes adorn the outer surface of the operculum and are a means of species identification.

Other unlikely-looking predators are anemones (Anthozoans), the exquisite flowers of the sea (Illustrations 45, 67–70). Closely related to corals, they share a similar body plan. While corals secrete a limestone cup to protect their vulnerable bodies, the anemones guard themselves with an awesome array of tentacles, armed with nematocysts. Any fish unwary enough to try a bite is soon enmeshed in a lethal web of filaments. In moments the fish is traumatized, then gradually transported to the central mouth for ingestion.

No one knows how long anemones live in nature, though one recorded specimen is said to have flourished in captivity for 90 years, dying only when a circulating pump failed and it was exposed to the killing air.

A number of other plankton-feeding species remain in one location to feed, but are not necessarily sessile and do not use armed tentacles in their feeding. The crinoids or feather stars (which we will treat at greater length in the next chapter) evolved in ancient seas as relatives of the starfish (Illustrations 61, 66). Rather than tentacles, the crinoid developed "arms" with a feathery array known as pinnules (Illustration 49). The pinnules are lined with tiny tube-feet, which catch passing plankton on a mucous coating. The plankton are then passed along the tube-foot of a pinnule to a food-groove which runs down each main arm to the central disc-like body. Some crinoid species develop a profusion of feeding arms, which, when fanned across the passing current, form a layered baffle. The water becomes trapped into eddies and swirls within the thicket of arms, allowing the crinoid to feed more effectively from the food stream. This process is easily demonstrated by releasing fluorescein dye upstream from the crinoid and watching the dye flow among the tangle of arms before escaping to the passing current. This feeding mechanism is so successful that these remarkable creatures are presumed to have used it essentially unchanged for some 400 million years.

Crinoids are not firmly rooted to their coral perch. They have a set of tiny legs, the cirri, with which they can climb to better feeding perches, or retreat out of the flow of current when they have finished feeding or when the current becomes too strong. Some South Pacific species can even be found in open water, swimming in the currents by lashing their feathered arms, perhaps seeking richer feeding grounds or else releasing gametes to spread their species across the sea.

Another important group of sessile plankton-feeders —among them sponges, and mollusks, such as oysters and scallops—have evolved a radically different method of extracting food from the waters. Using their bodies as filters, they ingest sea water, strain out their prey, and expel the water. The colorful sponges (Porifera) are a primary user of this system (Illustrations 51–53). We now know that a sponge is a single animal composed of a number of specialized cells working together to create an animated filter. Some of these sponge cells (collar cells) lining the interior body cavity possess flagella which beat continuously, creating a steady current through the animal's body. Other cells secrete stiffening limestone needles called spicules which help the sponge to hold its shape. Still others assimilate the filtered plankton and pass nutrients to the collar, epithelial, mesenchyme, and remaining specialized cells. A large sponge will pass hundreds, even thousands, of liters of water through its body in a single day. This is another phenomenon observable by introducing fluorescein dye into the ingested water and watching great clouds of colored water pour out of the body cavity. Some mollusks such as scallops, oysters, and Tridacna clams also use a flow of water through their sessile bodies to obtain nourishment. The filter feeding of both types is highly efficient. Divers have found complex or chambered sponges large enough to hold two persons. Tridacna clams can grow as large as 1.5 meters long and weigh hundreds of kilograms.

Methods of predation exist in profusion. Crabs, lobsters, crayfish, and shrimp walk around on spidery legs using their foremost claws to pluck tidbits from their surroundings. Many of these browsers prey on coral, algae, and smaller crustaceans. Other creatures that feed upon corals include the segmented bristleworms whose spun-glass bristles can give a bare-handed diver an unforgettable experience indeed (Illustration 54). Whenever a diver feeds fish, these voracious scavengers soon appear, swarming to clean up any fallen scraps.

Many species of fish hunt by swimming slowly through their world, peacefully grazing on coral polyps, algae, sponges, and other stationary feeders. Some of these, such as the butterflyfish, have even evolved long noses that allow them to reach into small crevices to feed. Other fish, such as the perpetually moving wrasses, dart from place to place, catching an unwary worm here, a shrimp there.

Still other reef fish are hovering plankton-eaters: for example, the earlier mentioned royal gramma and the jawfish. These two species bob and weave above protective shelters, snapping tiny bits of food from the passing waters, ready to dive for cover if predators threaten.

A large number of fish pursue their prey, using either blinding speed and/or maneuverability or both. They are guided by highly sensitive lateral-line detectors and their eyes. The swift jacks, barracuda, and sharks are all masters of the open water above the reef. Smaller fish enter that open water arena at their peril. The mortality rate is high.

Some fish, as someone once said, seem born merely to be eaten. Anchovy, silversides, and other "baitfish" are found in extensive schools hovering in the spacious water above the reef, often with jacks or barracuda making swift passes through the shimmering clouds of sleek bodies. To the diver and photographer there is in this encounter a kind of savage beauty. The silvery mass of fish molds itself like some magical fluid to the thrusts of the attackers. One may even see a predator plunge through a sudden tunnel in the shoal as the baitfish press away from danger. Then, as the predator passes through, the tunnel closes.

A different hunting strategy—ambush—must be used in close proximity to the reef structure. Because the prey is close to shelter in the labyrinth of hard coral, sweeping passes like those of the shark or amberjack would fail. Instead, a sizeable group of predators, such as groupers and sea bass (Serranidae), lurk unseen in the shadows of the coral structures. When unsuspecting prey swims by, the grouper charges forward with a flick of its massive tail. That booming tail-lash, which tolls a single requiem peal for the prey, rings out over the reef with startling effect. A diver will hear the boom, and see every fish in sight dart in unison to one side to avoid the charge. A moment later all is normal, and the reef populace resumes its serene pace. The price of serenity has been paid . . . until next time. There is a variation on ambush that might be described, "I'm sitting out here in the open, but you won't see me until I eat you." This technique is masterfully practiced by some of the more bizarre fish, including the scorpionfish, toadfish, lizardfish, and goosefish. Each of these sits motionless, camouflaged as a rock, a clump of algae, or a sponge, death-still eyes taking in every movement in its vicinity. When a target moves near, it is engulfed in a single explosive movement.

Each time I have observed a grouper, lizardfish, or other ambusher shoot forward and bite its prey, the victim seemed to suffer instant trauma. Once bitten, it never moved and was easily consumed by the predator.

One occasionally encounters fish swimming about with wounds in their sides, but these wounds are often signs of territorial aggression: punishing nips by their own or other species that thus forcibly make visitors to their range unwelcome. I can recall only one fish that had evidently been mouthed but managed to survive. This mauled four-eyed butterflyfish (Chaetodon capistratus) was observed swimming unsteadily about on Palancar Reef, off Cozumel, Mexico.

There are a number of predation specialties peculiar to certain families. The anglerfishes or frogfishes (Antennarius) have lures like tiny fishing poles mounted above their mouths. The frogfish can move these lures in a host of enticing ways, bringing small fish unwarily close to what seems like a bit of free-floating food. Then, snap! The frogfish opens its enormous maw so quickly that the prey is inhaled without the angler even changing position (Illustrations 55, 172).

Large snappers and groupers practice the same inhalation trick. I have fed groupers of ten-kilogram size from seven centimeters away and watched tiny chunks of food leap out of my fingers into the suddenly gaping maw, which instantly seemed to shape itself into a contented smile.

Predators are especially active at dawn and dusk, the feeding hours. This activity occurs in three distinct stages. Taking dusk as our example, the first change to be observed as daylight fades is the movement of vast numbers of diurnal fishes to their nocturnal shelters. Then, in the uneasy half-light, there is a period of half an hour or so, the so-called quiet period, which clearly favors the piscivores, or fish-eaters. Prowling in the gloom, these reflecting or neutrally colored predators such as sharks, barracuda, jacks, mackerel, tuna, and needlefish are nearly invisible as they approach, and their stalking strikes panic into the smaller fish (anchovies, herring, mullets) remaining in the open waters. This is the time when shoals of frantic fish roil the surface. Occasionally the charge of a tuna or mackerel will carry it one to two meters into the air. I have even seen sharks skip across the water for six meters or more in savage pursuit of panicky prey. Then within a half-hour the frenzied activity ceases as the predators can no longer see their prey, and great numbers of nocturnal fish rise from their reef shelters.

For the human observer, this twilight period is a rich and exciting experience, though sometimes fraught with hazard. I know of at least one diver attacked and killed by sharks because he went spearfishing at that hour. My judgment is that the diver would not have been attacked had he refrained from spearing at that time of day.

We should note that widely assorted predatory techniques have given rise to an equally impressive range of defensive responses. Intended victims, for example, can soar out of the water (flying fish) or form dense shoals (anchovies) from which the predators have difficulty isolating a target. Other potential victims hover near protective crevices or build burrows in the sand. Some have evolved long, sharp spines (sea urchins), while others have compounded sharp spines with an inflatable body (Illustration 60) to make themselves unappetizing to potential predators (porcupinefish and pufferfish).

As we shall see, still other forms of marine life become masters of color change and camouflage. Here we may merely note that species as diverse as squid, grouper, and flounder have control of their body color over a wide range of hues for both attack and defense. Corals pull inside their protective limestone cups; mollusks such as oysters, scallops, and clams slam shut their shells; small fish flee with the speed born of panic. Every reef creature has some kind of defensive system or technique.

To close this brief account of attack and defense, I can think of no better illustration of the two-edged sword of predation than my encounters with the small striped damselfish known as the sergeant major (*Abudefduf saxatilis*). The sergeant major is a perky, aggressive, nervously moving fish. When one brings a plastic bag of food into the water, this species can be counted upon to lead the pack coming to feed. In fact, there is a dive site off the Caribbean island of Grand Cayman where even carrying a plastic camera housing into the water can attract a swarm of sergeant majors.

In Hawaii I once found a spot where a sergeant major had deposited her eggs on a rocky surface. For some reason, the eggs had attracted several hungry butterflyfish, wrasses, and damselfish. These fish began attacking the eggs relentlessly, while the frantic sergeant major raced back and forth through the feeding frenzy trying to chase off the predators. The impulse to intervene, to somehow aid the mother rose within me. I even tried to wave the swarm away. Later it seemed to me that I had been the intruder, the extraneous element in that particular equation. Left alone, the reef system works. Feeding is the mainspring in the complicated society of the underwater wilderness. It guides all individual movement, shapes all species development. What we may see as a vicious attack, a tragic loss, has those attributes only in our eyes. In nature, such attacks and losses are merely isolated events in a brilliantly functional system which has assured the survival of incredible numbers of species.

45 *A large anemone*, Radianthus ritteri, *engulfing a squirrelfish*, Flammeo sammara. *Despite their delicate floral appearance, these large anemones use their poison-laden tentacles very efficiently to capture prey.* (*Rangiroa, French Polynesia*)

46. Spirobranchus giganteus, *a serpulid worm, removed from its limestone burrow. At the top are the crown of spiral feathers and the operculum or trapdoor with its intricate horns. This is a male and the smoky cloud to the left is its sperm. To the right are the eggs of a nearby female.*

47. *Stony coral polyps, growing on the mucus and sand tube which shelters the body of this sabellid worm. This is a rare occurrence.* (*Tahaa, French Polynesia*)

48. *A complex feeding arm of the crinoid* Nemaster rubiginosa. (*Bonaire, Caribbean*)

49. *A gorgonian.* (*New Caledonia, South Pacific*)

50. *A cluster of sabellid tube worms.* (*Cozumel, Mexico*)

51. *A vertical plate of the coral* Agaricia agaricites, *encrusted by a sclerosponge. Note the network of circulatory canals. The sponge's cells create currents of water from which nutrients are filtered.* (*Curaçao, Caribbean*)

52. *A tube sponge of the genus* Callyspongia. (*Cayman Islands, Caribbean*)

53. *The blood-red bucket sponge of the Cayman Islands. To the left are the polyps of the coral structure upon which the sponge has grown.* (*Caribbean*)

47

49

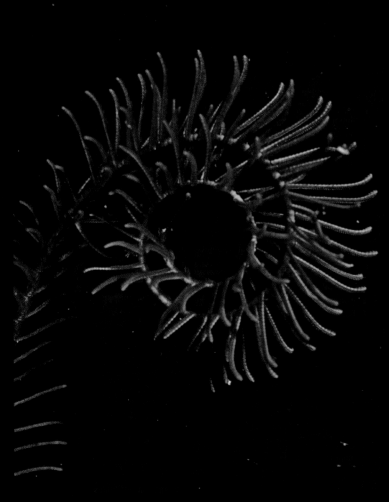

48

50

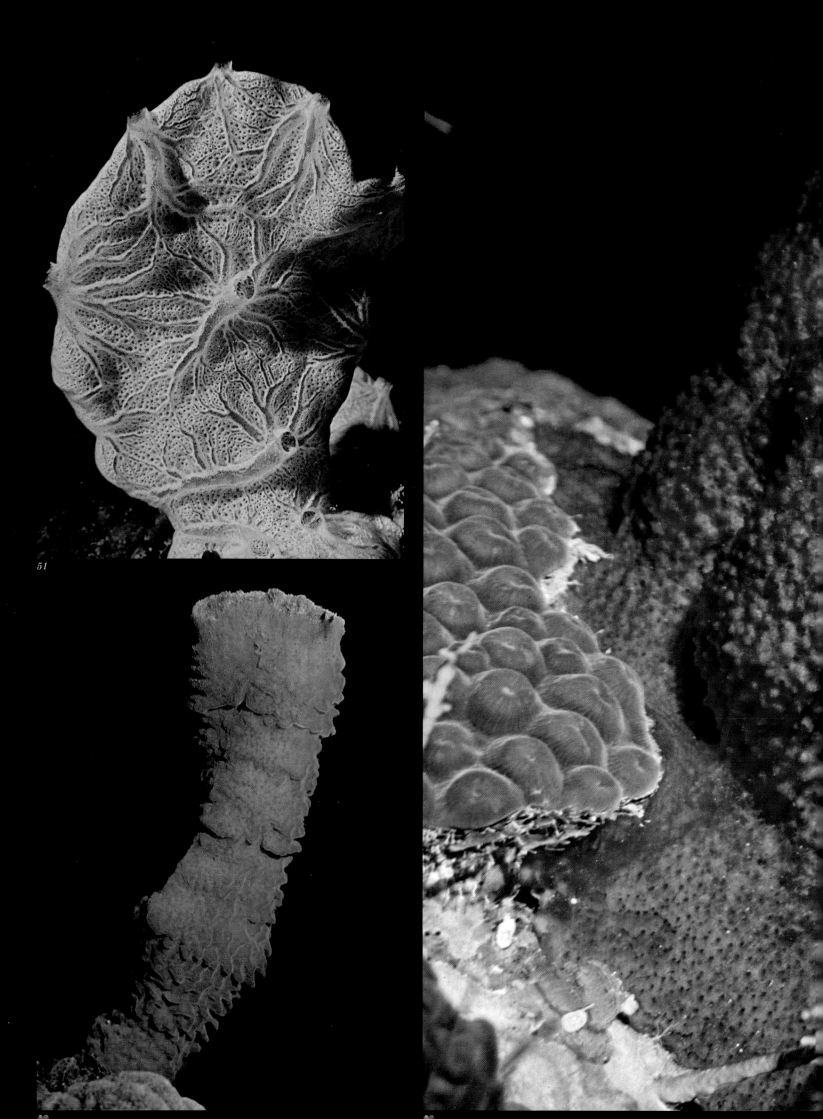

51

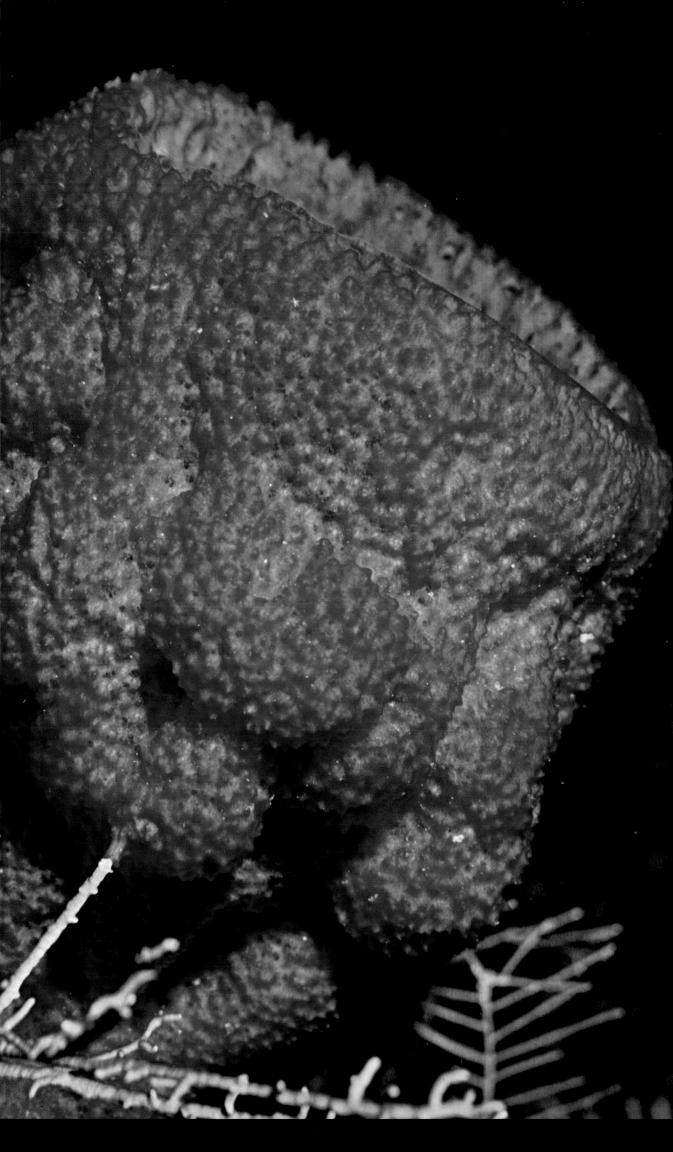

54. *A bristleworm feeding on polyps of the night-blooming coral* Tubastrea tenuilamellosa. (*Bonaire, Caribbean*)

55. *A longlure frogfish* (Antennarius multiocellatus) *sitting motionlessly between bread sponges (foreground) and a blue tube sponge. (Curaçao, Caribbean*)

56. *A sea urchin, often found on the lava boulders of the Sea of Cortez.*

57. Chromis caeruleus *retreating for shelter to an intricate thicket of coral branches.* (*French Polynesia*)

58. *A large Caribbean parrotfish,* Sparisoma viride, *using its armored beak to tear out chunks of coral, which it ingests.* (*Bonaire, Caribbean*)

59. *A rock hind* (Epinephelus adscencionis) *swallowing a live parrotfish. The prey's tail protrudes from the predator's mouth.* (*Bonaire, Caribbean*)

60. *An inflated spiny pufferfish* (Diodon holocanthus). *It thus becomes an unappetizing prospect to potential predators.* (*St. Maarten, Caribbean*)

56

57

58

59

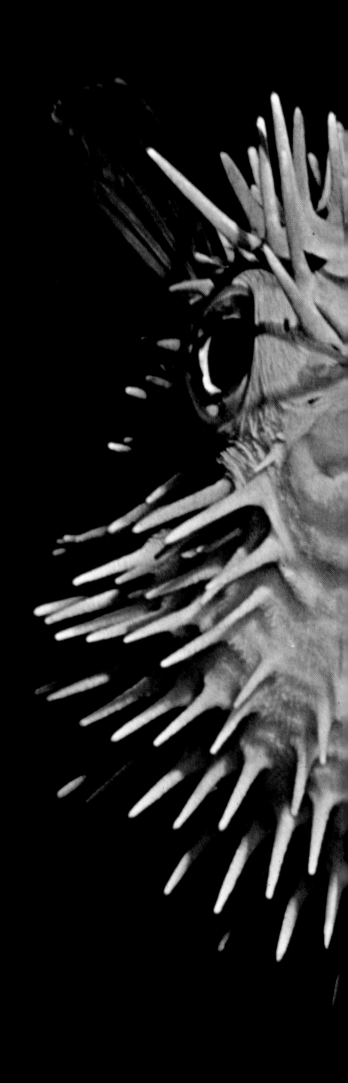

60

Cooperation and Aggression

Inhabitants of the underwater wilderness live in a complex society, marked by a variety of interrelationships. Predation is only one form of interaction, and a very brief one, lasting only as long as it takes the hunter to eat its prey. There are a number of continuing associations between forms of life in the sea. Many are lifelong affiliations involving two or more individuals in a colonial assemblage. We will examine five types of beneficial associations and then, in contrast, the rejection of association through aggression and territoriality. Despite popular misapprehensions, this rejection is not necessarily harmful or sinister. Indeed, we will see that such behavior plays a highly constructive role in the undersea society.

The closest continuing relationship is logically that of physiological symbiosis, in which there are two dissimilar organisms, neither of which can survive without the other. The term "symbiosis" originally was applied in the broad sense of "living together," but it has more recently come to be used in biology to signify a relationship involving some benefit to each member.

One example is that of the reef-building corals and the algae (zooxanthellae) imbedded in their tissue. Earlier I touched on the chemical benefit to the corals of the algae, which consume wastes and provide oxygen. In turn, the host provides the algae not only with a stable base in shallow water where sunlight penetrates, but also with carbon dioxide, nitrogen, and phosphorus for the chemistry of photosynthesis. Recent research has shown that the relationship goes even further. It now appears that certain such partnerships survive in extended darkness through consumption by coral of a scale-like substance secreted by the zooxanthellae. This would explain how some soft corals have practically abandoned the capture of living prey, receiving all their nutrients from their symbionts.

The zooxanthellae establish a different relationship with another of their hosts, the Tridacna clam. The lush mantle of these large bivalves contains immense concentrations of algae, due to lens-like elements on the outer layer of the mantle which focus light deep into the host's tissue (Illustration 229). Unlike the coral-zooxanthellae symbiosis, however, this one has a definite ending: the clams eventually digest their tiny partners.

A partnership called *inquilinism* exists between the small damselfish or clownfish (*Amphiprion*) and large *Stoichactis* anemones in the Indo-Pacific (Illustrations 67–70). In inquilinism, one partner seeks refuge in the other. The inquilinism partnership of these two families is complex and stems from a long evolutionary adaptation. The large anemones are armed with a multitude of tentacles equipped with lethal nematocysts that can engulf a fish several centimeters long with impressive dispatch. Yet the clownfish blithely swims among the tentacles, rub-

Modified dorsal fin

A shark and a sharksucker or remora. The shark-sucker's anterior dorsal fin forms a suction cup that enables it to attach itself to its large host.

bing his body against them frequently as if to say, "It's only me."

Research has shown that the clownfish and the anemone go through a specific ritual in "adopting" each other. In this ritual the clownfish repeatedly brushes the virulent tentacles in a familiarization gesture. Soon the clownfish has stimulated its own epidermal cells to produce a chemical mucus that does not activate the powerful nematocysts even in continuous rubbing. The activation factor is critical, however, for clownfish wiped clean of mucus have been devoured by their host anemone.

The inquiline relationship may also be mutualistic. It is clear that the clownfish draws protection from its association with the anemone, but the anemone's benefit, if any, is not well understood. There is the fact that clownfish have taken scraps of food from divers back to their anemone. The anemone then captured the food from the clownfish. The clownfish may also act as a lure, drawing, with its brilliant colors, would-be predators that are then enmeshed by the anemone. The fact that the clownfish hover and dance above the anemone and withdraw into its embrace when approached may well signify such a service. They may also clean debris and wastes from their host.

We find a parallel relationship in the Caribbean (where there are no clownfish) between the shrimp *Periclemenes yucatanicus* and *Condylactus* anemones (Illustrations 75–76). Other species share inquiline or refuge associations. These include the small man-of-war fish, found only among the dangerous tentacles of the Portuguese man-of-war (*Physalia*), and a variety of small fish that shelter among the needle-spines of the sea urchin *Diadema*.

Another type of interspecies association is *endoecism*, in which two animals share a burrow or tube. A dramatic example of this is the crustacean-fish relationship typified by the prawn-goby pairings of the Indo-Pacific. On sandy bottoms throughout the islands one can find open burrows, each guarded by a watchful goby five to seven centimeters long. If you take up a position a few meters away, you will soon see a shrimp emerge from the burrow carrying sand or stones in its pincers. Dropping the debris well clear of the mutual burrow, the busy shrimp hustles back for another load. When disturbed, both animals retreat into the burrow. The shrimp, which has extremely poor eyesight, is alerted by the goby when danger approaches. The crustacean never resumes working until the goby again stands watch.

A fourth form of association between marine species is *epizoitism*, wherein members of one species attach themselves to members of other species. Epizoitism ranges from parasitism to mere preference. In all of these cases it appears that the host offers a precisely defined micro-environment to the epizoite. It also becomes clear that many epizoites have developed remarkable powers of discrimination. The evidence now seems to support the idea that there is recognition at the molecular level. Thus, an epizoite responds chemically, say, to a certain protein in the outer shell of its host.

Examples of epizoitism are legion. They include the parasitic isopods and copepods (Illustrations 84, 96) which plague various marine hosts, brittle starfish adapted to certain hard or soft corals, and hydroids of several species whose preferences range from breadsponges in the Caribbean to certain kinds of scorpionfishes in the Indian Ocean.

A fifth category of association is *phoresis*, the relationship of transport, in which one organism habitually allows itself to be carried about by another. Among the fish that engage in phoresis are the echeneids, which include the remora and sharksuckers. These unusual fish range from a few centimeters to one meter in length, and have a scythe-swing swimming motion. Their peculiar ability to hitch a ride is due to the evolution of their anterior dorsal fin into an oval, ribbed suction cup with an edge that can be raised. By use of this remarkable device the remoras create a suction adhesion and are towed passively by their large hosts. Some species of echeneids seem to prefer the hulls of ships, which can give a long ride indeed. A few species of echeneids are known to provide grooming or cleaning services for their partners, but in most cases the host seems to derive little benefit from the transport association.

Other phoretic combinations include a wide range of barnacles on hosts such as *Mola mola* (the ocean sunfish) or whales, and sponges which are carried by certain crabs to mask the tasty bodies of the crustaceans. In some cases the dispersal of the rider may be important to the survival of its own kind, particularly in sedentary species. But generally this is not a crucial aspect of phoretic associations.

Perhaps the most prevalent interspecies association is *commensalism*, or "at table together," in which partners share food sources, generally with no harm to each other. One partner usually obtains the food, thus defining the other as a dependent but not parasitic guest. Among the organisms that provide fascinating examples of commensalism are certain types of anemones which attach themselves to the shells of browsing mollusks. Other kinds of anemones attach to branches of gorgonian (coral or coral-like organisms) to elevate themselves to a more advantageous site in the food stream. Some large sponges pass great volumes of food-bearing water through their bodies and thus become a haven for commensals. Over 16,000 shrimp were taken from one large loggerhead sponge in the Caribbean.

A uniquely interesting commensal association is that of the brilliant crinoids of the remote reefs of the Coral Sea (Illustrations 250–252, 258). The Coral Sea crinoids were the most spectacular I have seen anywhere, looking as if they were colored by a mad artist with an infinite palette. They were each inhabited by up to several hundred commensals ranging from shrimp and copepods to galatheid crabs (Illustrations 74, 78). In each case these polychromic commensals had adopted colors that specifically matched their crinoid host, and they were almost invisible among the crinoids' tangled arms.

Of all this swarm, the most distinctive individual (Illustrations 77, 254) was the clingfish (*Lepadichthys lineatus*). These exquisite, tiny fish resemble the gobies of the Caribbean known as cleaner gobies, but are distinguished by their polychromism, their ventral sucker disc, and a "neck." Observations by scientists on the morphology of the clingfish show that while attached to a surface by its disc, it can move its head remarkably. Unlike any other fish, the clingfish can raise or lower its head 25 to 35 degrees, move it sideways, and even rotate it eight to ten degrees. All this is made possible by unique vertebrae, long, cylindrical portions of which fit into grooves.

While I have presented several examples of association, each in its scientifically isolated aspect, nature is not nearly so tidy. In most cases, associative relationships involve two or more of the types I have described: transport plus sharing food, or sharing a burrow plus offering some mutual benefit, such as defense. Contrasted with associative relationships are those which are exclusive, typified by displays of aggression, hostility, and territoriality.

In the same way that symbiotic and commensal relationships have strong survival value, aggressive and territorial behavior fulfills important functions in the undersea society. Konrad Lorenz, in his seminal work *On Aggression*, reported extensive observations of interspecies and intraspecies aggression among tropical reef fish. While his analyses were stations on his way to conclusions regarding human society, his observations on reef fish were in themselves revelations about the underwater wilderness. Lorenz' studies centered upon particular phenomena: how territories were established and what caused their boundaries to shift, and the function of brilliant coloration in what he defined as "poster-colored" fish, such as butterflyfish, angelfish, and demoiselles. Lorenz discovered that territorial lines were generally a function of intraspecies aggression and served a survival purpose. They kept members of a single species from overpopulating a biota, assuring that the area would support the territorial inhabitant in its feeding specialty. Thus, a given range, or individual living space, could support one grouper; a smaller, perhaps overlapping range one damselfish; an entirely different range one pufferfish, and so on, just as a small town often supports only one doctor, one plumber, one auto mechanic.

The aggression that Lorenz recorded on the reef was found to have at least three vital functions: balanced distribution of animals of the same species over the available environment, selection of the strongest breeding stock by fights between rivals, and defense of the young. Similarly, he recorded two species-preserving functions of the poster colors on reef fish: the colors elicit a furious territorial defense in every fish of the same species in its own territory, and they proclaim a readiness to fight a foreign intruder. When combined with ritualized, nonlethal forms of combat, these two functions spread the population optimally over the available reef range.

Other types of territorial behavior occur also among dull-colored fish. In these cases we are generally observing territorial aggression within a species, such as that of the mouse-brown dusky damselfish (*Eupomacentrus dorsopunicans*), which attacks and chases far larger parrotfish, angelfish, and other large intruders and even nips divers. It is disturbing to be taking close-up photographs of coral and find your scalp being repeatedly nipped by a fish, even though it is only five centimeters long. Waving your hand violently at the damselfish doesn't stop it; it tries to nip the hand.

In this chapter we have juxtaposed highly cooperative behavior with aggressive behavior to show that they both play important roles in the mosaic of underwater life. Along the way, we also note that the more closely we observe such a society the more intricate and interrelated the associations of its individual members appear.

61. *A crinoid positioned on a gorgonian sea fan, exposing itself to the food-bearing currents. In this attitude it competes directly with the gorgonian for food.* (*Australia*)

62. *A colorful crinoid from the Coral Sea.* (*Australia*)

63. *A crinoid with its arms spread for feeding.* (*Australia*)

64. *The crinoid* Nemaster grandis. (*Caribbean*)

65. *A crinoid after feeding, with its arms folded. This often occurs after a strong current has provided rich feeding for the animal.* (*Australia*)

66. *A type of crinoid frequently found atop shallow coral reefs in tropical Australia.*

67. *A large anemone,* Radianthus ritteri, *completely withdrawn while feeding. In this position, it leaves its associated clownfish,* Amphiprion perideraion, *exposed.* (*Australia*)

68. *The clownfish* Amphiprion melanopus *amid anemone tentacles and coral polyps.* (*Australia*)

69. *The clownfish* Amphiprion tricinctus *amid the tentacles of the carpet anemone* Stoichactis giganteum. (*Australia*)

70. *The clownfish* Amphiprion tricinctus *in the anemone* Physobrachia douglasi. (*Truk, Micronesia*)

62

63

64   65

68

69

70

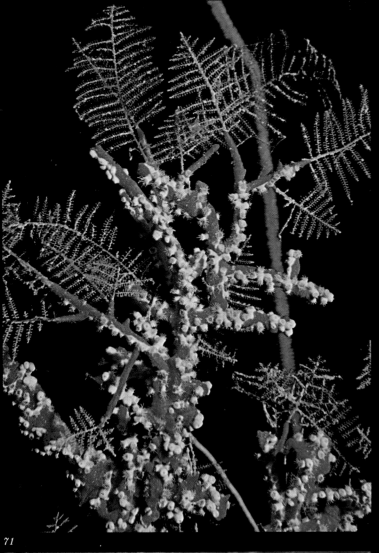

71

73

72

74

71. A dense Caribbean community: sponge and epizoanthids living on the skeleton of a hydroid colony.
72. Epizoanthids and sponges, often found together in Caribbean reef communities. (Cozumel, Mexico)
73. Long-snouted hawkfish (Oxycirrhites typus). (Palau, Micronesia)
74. The polychromic galatheid crab (Allogalathea elegans). The color of each of these crabs matches that of the crinoid on which it lives. (Australia)
75. The shrimp Periclemenes yucatanicus, which lives out its life amid the protective tentacles of an anemone. (Curaçao, Caribbean)
76. Another specimen of Periclemenes yucatanicus. It blends well with the purple coloration of its host anemone. (Curaçao, Caribbean)
77. The polychromic clingfish (Lepadichthys lineatus) on the feet, or cirri, of an inverted crinoid. It is the only fish with a movable neck. This is due to the unique vertebrae in the neck. (Australia)
78. A polychromic shrimp giving birth among the arms of its host crinoid, an event rarely recorded. (Australia)

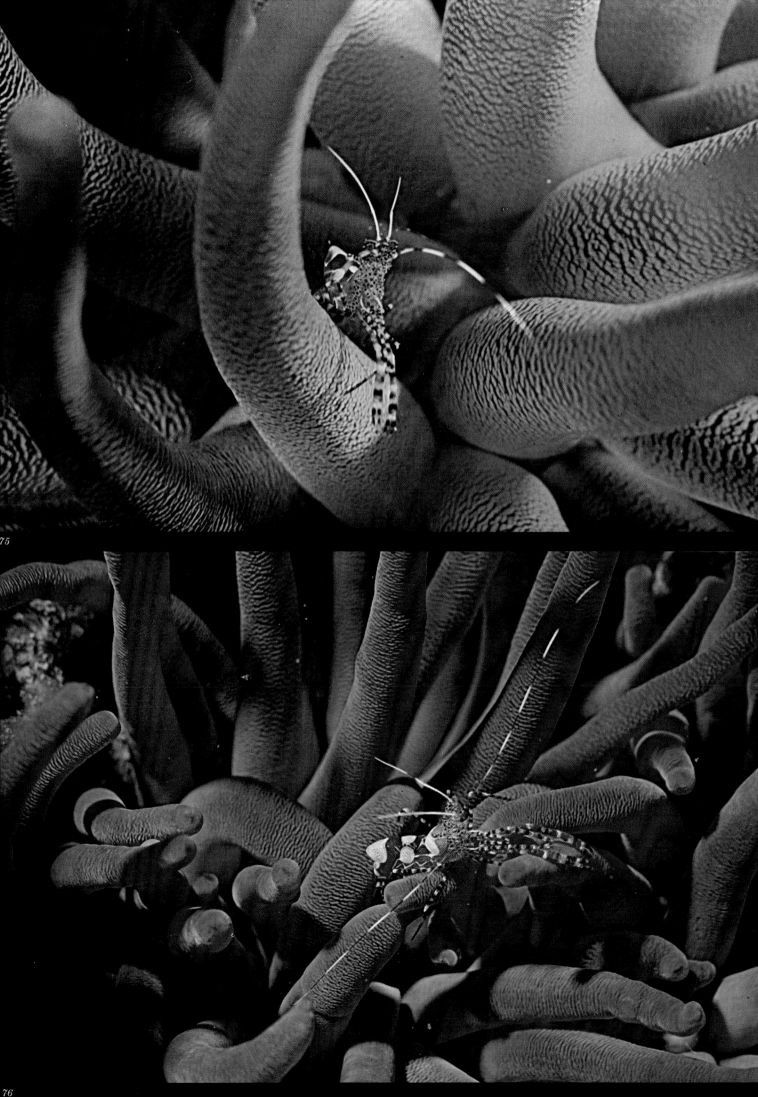

75

76

77

78

Parasites versus Cleaners

Two important types of association were omitted from the preceding chapter so they could receive expanded treatment here. This unique pair is parasitism and its counter, the cleaning function. In parasitism, an organism lives and feeds permanently or temporarily upon the body tissue of a host of another species, obtaining meals from the host's body but not destroying it. Continuous unchecked parasitism often does lead to the debilitation and death of the host, but in many forms of parasitism, the harm is not so severe that the host is lost.

In the cleaning function, certain species remove parasites and perform other vital grooming services for their hosts. The wonder of this behavior is that the parasite-plagued host is usually a large predator, such as a moray eel or grouper, many times the size of the cleaner and potentially a great danger to it. Yet within certain boundaries the cleaner pursues his services with total immunity. More than 40 species of fish are now recognized as cleaners, with at lease six species of shrimp and one species of crab known to provide similar services. More and more divers using scuba equipment are witnessing this intriguing ritual of the reef.

How does parasitism come about? In part, the answer is that certain of the close, lifelong relationships we discussed in the last chapter evolved along the ominous pathway of parasitism. After all, it is not so long a step from sharing a burrow or a shell to fancying the tissue of one's host. The Mediterranean pearl fish, which inhabits the anal cavity of certain large sea cucumbers, seems today to be at this evolutionary crossroads: specimens of such fish have been found that not only seek shelter in, but nibble the gonads of, their host.

Among the most interesting parasites are isopods and copepods, orders of free-living crustaceans many of which have developed specialized mouth parts capable of piercing the skin of a host. These parasites infest not only the outer skin but also frequently enter the body cavity through the mouth and gill covers. On some hosts, such as the Caribbean creolefish (*Paranthias furcifer*) and blackbar soldierfish (*Myripristis jacobus*), one may see large specimens of isopods half as long as the host's head (Illustrations 84, 96). I have watched other species, such as the tiger grouper (*Mycteroperca tigris*) and graysby (*Petrometopon cruentatum*), and seen almost invisible crustaceans skitter continually over their bodies. In response to this phenomenon, certain species have developed over a long span of time into cleaners, that is, creatures who prey on the parasites they find on the host. When one observes tiny shrimp moving unconcernedly over the body of a huge moray eel (Illustration 92), or small gobies in the open mouth of a large grouper, one must admire the amazing precision of the rules governing this encounter. For cleaning is without question highly organized and presumably advantageous to both parties.

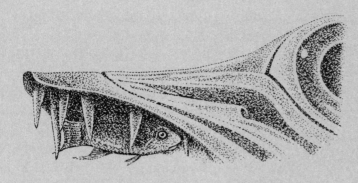

A Spanish hogfish has entered the mouth of a barracuda, usually its enemy, to feed on the parasites that afflict the huge fish. Thus both benefit from this symbiotic relationship.

Some of the most common cleaners I have observed at work are the Caribbean cleaning gobies (*Gobiosoma evelynae, G. louisae,* and *G. genie*); their Indo-Pacific counterpart the wrasse *Labroides dimidiatus;* the Baja California striped butterflyfish *Heniochus nigrirostris* (known as "El Barbero" or "the barber" because of its services); the shrimps *Hippolysmata grabhami, Stenopus hispidus,* and *Periclemenes pedersoni;* and juveniles of the French angelfish (*Pomacanthus paru*), gray angelfish (*P. arcuatus*), and Spanish hogfish (*Bodianus rufus*).

Patterns that appear quite significant recur among the activities of these cleaners and their hosts. For instance, the cleaners tend to inhabit certain areas at all times, establishing "cleaning stations." Caribbean gobies sit upon certain sponges or coral heads, while shrimps inhabit certain crevices, sponges, or anemones. Day after day a human visitor will find the same shrimp in the same anemone, the same gobies on the same coral head. The cleaner is thus assured of a regular food supply (though most do have standby sources) as parasite-ridden hosts queue up to be cleaned. In return, the hosts know where to apply for relief from parasite infestation.

A second recurrent pattern in cleaners is their coloration. Caribbean gobies, Pacific cleaner wrasses, cleaning shrimps, and the cleaning juveniles mentioned above are so often striped that there seems to be a connection between this color pattern and this activity. In fact, when French angelfish grow to adulthood and cease functioning as cleaners, they lose their distinctive juvenile pattern of bold stripes (Illustrations 98–100). Also, it is now known that certain remoras and sharksuckers (some of which have strong body stripes) offer cleaning services to sharks and rays.

Even stronger evidence has recently come from experiments in which Caribbean groupers, serviced by Caribbean cleaning gobies, were placed in aquaria with Pacific cleaner wrasses (*L. dimidiatus*). There is a similarity in color between these gobies and wrasses, though the wrasse is considerably larger and moves quite differently. After a brief period of confusion the groupers did allow these completely strange, but striped, fish the universal recognition extended to cleaners. Such evidence, while not conclusive, seems to support the theory that stripes are often a sign of a cleaner fish.

Many cleaners also have certain attention-getting movements which advertise their whereabouts and availability. Barbershop shrimp hide their vulnerable bodies in crevices but jiggle their long white antennae out in the open (Illustrations 81,85,89). The Pacific cleaner wrasses do a jiggling dance that is unmistakable, almost hypnotic. The Pederson shrimp sits on its protective anemone and shudders conspicuously (Illustration 93).

Once these signals have been received by the host, the most fascinating behavorial sequence to be seen under

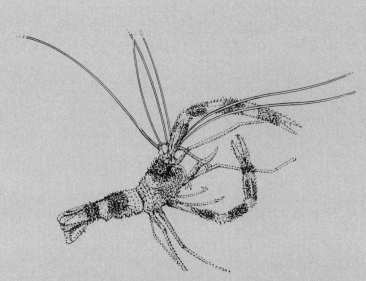

Cleaner shrimp. These tiny creatures habitually pick parasites from the bodies of large fish that welcome the service.

the sea follows. Cleaner and host engage in an elaborate ritual whose fairly evident purpose is to offer the cleaner immunity from predation if it will relieve the host of its parasites. This ritual varies in its specific gestures from host to host and cleaner to cleaner.

The "acceptance" behavior of the host is also highly variable between species. Small host species such as bonnetmouths (*Emmelichthyops atlanticus*), creole wrasses (*Clepticus parrai*), and various parrotfishes (*Scarus* species) hover near the cleaning station in what are extraordinarily atypical poses for fish. They hold themselves head up or head down in orientations which evidently take a good deal of effort to maintain. In some cases this posture is accompanied by holding the mouth fully open throughout the ritual. Other, larger predators, such as moray eels, groupers, and jacks, glide up to the cleaning station, and hang there motionlessly. The "cleaner's acceptance" follows, with the cleaner swimming or hopping from its perch onto the body of the host. Moving in an orderly fashion from spot to spot, the cleaner not only eats parasites but also sometimes removes infected scales and even seems to treat the edges of wounds. Since many parasites freely move into the host's mouth and gills, the cleaners follow, swimming confidently into a mouth held obligingly open often for minutes at a time.

The process ends either with the host being satisfied or interrupted. Having caused such an interruption many times, I have seen certain characteristic signals which tell the cleaner unmistakably to leave. These are usually given only by the larger eels and groupers, since many other host fish simply bolt and leave the cleaner hanging. When approached by a diver, the groupers and eels give a visible shudder, which is immediately followed by all the cleaners returning to their protective stations. The grouper or moray may not leave, but it is ready for whatever action becomes necessary.

This entire ritual is a kind of negotiated truce. Only during this truce is the cleaner safe from predation by the host. At any other time most of the cleaners are considered wonderfully appetizing morsels by the same hosts who so diplomatically welcome them during the cleaning ritual. So much for diplomacy. I once extracted a number of cleaner shrimp from their protective crevices and photographed them falling free in open water. When released, they would head for the protective reef as quickly as they could. I soon learned why. Away from their protective, well-known cleaning stations, they are quickly snatched and eaten by the nearest graysby, coney, or grouper, which always seem hungry for a crustacean snack. Thereafter I escorted all photographic models home after their assignments, and the mortality rate quickly dropped.

In another remarkable glimpse into the endless workings of evolutionary forces, there is at least one documented case of creatures who thwarted the cleaning relationship for their own purposes. Irenaus Eibl-Eibesfeldt, Wolfgang Wickler, and other behavioral scientists have reported an amazing look-alike for the Pacific cleaner wrasse. This mimic, the saber-toothed blenny (*Aspidontus taeniatus*), is almost identical in coloration to that wrasse. The saber-toothed blenny carries its imitation so far that it even matches minor local color variations in the cleaner. Not only is the blenny similarly colored, but it also stays in one area, frequently near a cleaning station, and does an inviting dance very like that of the wrasse. All of this activity supports the saber-toothed blenny's diet, which consists of fish skin and scales. Since fish are not normally willing to have their scales bitten off, this mimic has evolved its feeding technique under the ornate cover of the cleaning ritual. When it has lulled a fish into exposing itself, the blenny darts in and bites a chunk out of the host. Observers have noted that these bites cause the hosts visible distress.

Older fish who have suffered repeated attacks, however, seem to perceive the differences between the cleaner and the mimic. For this reason, *Aspidontus taeniatus* is most successful when it puts on its show for younger, less experienced fish. Indeed, it has been reliably reported that cleaner mimics in aquaria have been abruptly eaten in the middle of their show by perceptive groupers. After several days together in captivity, the groupers seemed to recognize the mimic's act and terminated it with one gulp.

Before leaving the visually rewarding behavior of cleaning, we should ask whether it is a mere curiosity, or whether it performs a vital function. This question was anticipated some years ago by a fine researcher, the late Conrad Limbaugh, whose death in a cave-diving accident cost marine biology an inquiring and original mind.

Limbaugh isolated a series of patch reefs and carefully removed from them all species known to render cleaning services. The results were dramatic, the conclusions clear. Within two weeks many species had forsaken the parasite-ridden sites to take up residence elsewhere. Among the species that stayed, an epidemic of frayed fins, sores, and fungus infections eroded the general level of health to the point where the survival of the inhabitants was threatened.

79. *Cleaner wrasse* (Labroides phthirophagus) *with saddleback wrasse* (Thalassoma duperreyi). *(Hawaii)*

80. *Two small cleaner gobies* (Gobiosoma genie). *(Caribbean)*

81. *The barbershop shrimp* (Stenopus hispidus), *found in most of the world's tropic seas. It provides cleaning services to large predators, such as moray eels and groupers.*

82. *Cleaner goby on the red hind* (Epinephelus guttatus). *(Saba, Caribbean)*

83. *A cleaner goby* (Gobiosoma genie) *servicing a tiger grouper* (Mycteroperca tigris). *The goby not only cleans the body of its client, but also swims into its mouth and gills in search of parasites. (Bonaire, Caribbean)*

84. *Parasitic isopod on forehead of* Myripristis jacobus, *the blackbar soldierfish. The same phenomenon, photographed at night, is shown in illustration 96. (Cayman Islands, Caribbean)*

85. Stenopus hispidus *perched in a tube sponge. (Curaçao, Caribbean)*

86. *Cleaner wrasse* Labroides dimidiatus *with the pufferfish* Arothron nigropunctatus. *(Red Sea)*

87. *Nassau grouper* (Epinephelus striatus) *being cleaned by* Gobiosoma genie. *(Cayman Islands, Caribbean)*

88. *Cleaner gobies* (Gobiosoma genie) *using a colorful sponge as a cleaning station. (Cayman Islands, Caribbean)*

89. Stenopus hispidus *forced from its protective crevice into a vulnerable location atop a coral head. Left alone, it will rush back or risk being eaten. (Bonaire, Caribbean)*

90. *Cleaner gobies* (Gobiosoma genie), *dwarfed by their host, a green moray eel* (Gymnothorax funebris). *(Curaçao, Caribbean)*

91. *Cleaner gobies on Nassau grouper. (Cayman Islands, Caribbean)*

92. *A spotted moray eel* (Gymnothorax moringa) *with two cleaner shrimps* (Stenopus hispidus). *(Bonaire, Caribbean)*

93. *The cleaner shrimp* (Periclemenes pedersoni). *(Curaçao, Caribbean)*

94. *An enormous green moray* (Gymnothorax funebris), *nearly three meters long. Its body was stained with rust from the wreckage of the sunken ship in which it lives. Tiny cleaner gobies search its body for parasites. (Bonaire, Caribbean)*

95. *Cleaner wrasse* (Labroides phthirophagus) *cleaning gray wrasse* (Thalassoma ballieui). *(Kona Coast, Hawaii)*

96. Myripristis jacobus, *the blackbar soldierfish, plagued by a large parasitic isopod on its forehead. On Caribbean reefs it is unusual for the parasite to be so large in comparison to the host. The same phenomenon observed by day is shown in illustration 81. (Cayman Islands, Caribbean)*

97. *A juvenile French angelfish* (Pomacanthus paru) *offering cleaning services to a school of goatfish* (Mulloidichthys martinicus). *One of the goatfish is changing color to request cleaning. (Curaçao, Caribbean)*

98. *A juvenile French angelfish* (Pomacanthus paru). *(Curaçao, Caribbean)*

99. *The changing colors of an adolescent French angelfish. In this transition period it still retains its juvenile striped pattern while developing adult scale markings. During the loss of body striping, the fish ceases to offer cleaning services. (Bonaire, Caribbean)*

100. *The adult French angelfish with distinctive edge markings on the body scales. The juvenile striped pattern has disappeared. (Cayman Islands, Caribbean)*

80

84

81

85

82

86

83

87

88

92

89

93

90

94

91

95

98

99

Color Magic

The observer of the underwater wilderness is inevitably struck by the lavish use of body colors in a virtually monochromatic world. Marine life comes in a bewildering array of colors and patterns, each seeming to suit some purpose of survival. The effects range from the muted gray flannels of the sharks to the gaudy finery of the angelfish or lionfish. These effects are inevitably of great benefit. All marine coloration seems to be of three basic types: colors for a lifetime, those that change during a lifespan, and those that can change every day.

Most sharks are stealthy hunters of the open water. For them the dress most effective in the hunt is *obliterative countershading*. Their upper bodies are darker, to blend into the dark water when they are seen from above; their bellies are light, to blend into the sky when seen from below. When the sunshine lightens their backs and their body shadow darkens their undersides, these ghostly animals simply disappear into the void of their background, awaiting a chance to strike. Since their mode of predation remains unchanged throughout their lives, their body colors remain constant as well.

Another kind of lifelong, purposeful attire is *disruptive coloration*, in which strong stripes or other markings break up the normal fish-shaped body outline and render the fish less distinctly visible to predators when seen against the broken, angular contours of the reef. This kind of coloration is shared by such diverse fishes as snappers, drums, and hawkfish. The "poster-colored" fish is so gaudy it could become a tempting target for predators if it did not also make use of certain *directive* or *deflective* markings. Scientists have noted that marine predators often aim for the eye of their prey. If one or both eyes are put out of commission, the victim has little chance to survive on a reef. As a defense, numerous species, such as the butterflyfish, have evolved resplendent "eyes" far from their true eyes (Illustrations 17,25). These and other species also tend to disguise the location of their true eyes with strong "eye-stripes" (Illustrations 13,20,22,24). Others, such as the imperial angelfish (*Pomacanthus imperator*) and the masked butterflyfish (*Chaetodon lunula*), render their eyes practically invisible through massive swatches of jet black on their faces (Illustrations 16,19,23). Many predators seem to assume that the prey's body mass will follow its eye in flight. Occasionally, a charging attacker may aim for the false eye. Expecting the prey to flee in the direction indicated by the false eye, the predator may overshoot the prey as it flees in the opposite direction. This deflective coloration is an effective defense, for the prey protects its true eye by drawing a charge at its less vulnerable tail. Other species can fade into relative invisibility by being nearly transparent, or, among deeper-dwelling reef animals, by being jet black or deep red. In the darkness of the deeper slopes, a black fish is virtually invisible.

Many nocturnal species, such as the squirrelfish, cardinal fish, and bigeye, hunt in moonlight and have developed red coloration (Illustrations 142, 152). Light is selectively filtered as it passes through the upper waters of the sea. In this selective light absorption, the red end of the spectrum is the first to be filtered. Thus, a red fish at a depth of 50 meters or more becomes practically colorless and effectively disguised. For these nocturnal hunters, the fact that even the red component of strong moonlight will not reach them means they can move about and feed in relative safety.

Another survival ploy involving basic body colors is *mimicry*. Predators are often pigmented to resemble something harmless, while their prey is often attired to resemble something inedible. For example, the lethal stonefish and scorpionfish have an algae-like fringe below their mouth. This fringe serves to break up their body outline and camouflages the ever-waiting mouth. The varicolored frogfish can sometimes be mistaken for nearby yellow, red, or brown sponges. I have seen black frogfish sit invisibly just inside the shadowed body cavity of a tube sponge. Defensively, some small fish will resemble a drifting leaf, or as in the case of the sargassumfish, floating seaweed.

All of these animals are born with a certain coloration that suits their habits. These animals keep the same coloration and the same predation techniques throughout their lives. But there are more complex uses of body color. While some of these have obvious purposes, others are imperfectly understood. Strongly striped juveniles of French angelfish and gray angelfish act as cleaners, yet cease that activity when, unstriped, they achieve adulthood. The behavior and the coloration are consistent. There is a strong temptation to apply this fledgling hypothesis to other species whose coloration appears similar. But the undersea society will not let us cubbyhole it so easily; the rules are more complex than we have perceived.

Juvenile spotted drum and jackknife fish (*Equetus*) are strongly striped and yet do not appear to offer cleaning services (Illustration 175). In this case the stripes appear to be merely disruptive coloration. It may be significant that these fish retain their stripes as adults. Still, one would expect these distinctively striped juveniles to offer cleaning services. Yet, they do not.

Two other aspects of body coloration are highly unusual in that they involve voluntary color change controlled by the nervous system. The colors change from hour to hour or moment to moment in response to the subject's emotional states—a great advantage in both offense and defense.

One situation in which the ability to change color plays a significant role is the cleaning ritual. On a reef in Curaçao one day I observed a school of yellow-tailed goatfish (*Mulloidichthys martinicus*) glide up to a small coral head, the cleaning station of a juvenile French angelfish. To my surprise, one goatfish abruptly flushed a deep red, whereupon the tiny cleaner came and serviced it (Illustration 97). In my eagerness to photograph this incredible moment I came too close, and the school of goatfish swam away. Even as I sat back in disappointment, my eyes followed the school, particularly the one goatfish which had resumed its normal yellow and white body colors. My stillness and retreat reduced the disturbing effect of my presence, and as I watched spellbound the school slowly eased back to the cleaning station and the same goatfish flushed red once more. Having held my breath for what seemed hours, I took a picture. My explosive exhalation again disturbed the goatfish, and they swam off. This was repeated eight times, and in every case only that same goatfish flushed red at the station and received cleaning from the young angelfish. Was the coloration a request, a truce sign? Or did it perhaps render the parasites more easily visible? Much remains to be learned about these remarkable phenomena.

Another family of fish with extensive color control is that of the Serranidae (sea basses, groupers, coneys), many of which display two completely different color control patterns, one related purely to age, the other to adult emotional stimuli. Take tiger groupers (*Mycteroperca tigris*) as an example. The juveniles are bright yellow with a brown lateral stripe (Illustration 125); adolescents have vertical white stripes on a yellow body (Illustrations 111, 126), and the normal coloration of adults is a brown, tiger-striped body (Illustration 112). We can only speculate why. Perhaps the brightly hued juveniles thus escape predation by adults of their own species. In a completely distinct phenomenon, adult tiger groupers have a certain repertoire of extemporaneous color control, flushing nearly white, deep brown, or shades of roan red. The very dark coloration seems well adapted to lurking in shadows, rendering these groupers more effective as ambushers. I have observed the white coloration both as camouflage over a white sand bottom and as part of what seemed a courting or territorial ritual. In this encounter, two large groupers swam in a tight circle, nose to tail, with one flushed dark and the other light. Since I intruded to photograph the event, I could never ascertain the normal course or outcome. A law of physics states that certain phenomena cannot be observed because the act of observation alters the occurrence. The limited range of underwater lights and camera optics has made me subject to this excruciating principle on all too many occasions.

The coney (*Cephalopholis fulva*) is another serranid which has two distinct adult color variations in which the fish's emotional state evidently plays a notable role. When observed from a distance, adult coneys tend to shades of brown with ocellated (eye-like), black-rimmed blue spots. As the observer draws near,

Color changes in certain marine fishes and invertebrates are brought about by the expansion or contraction of pigments in the chromatophores (pigment-bearing cells).

The black and stippled pigments are contracted. Here, as below, the chromatophores are surrounded by cells with a white or iridescent aspect.

The black pigment is expanded; the stippled pigment is contracted

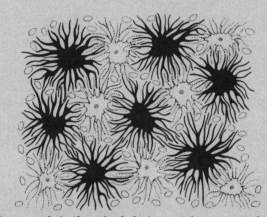

The black pigment is contracted; the stippled pigment is expanded

Both the black and the stippled pigments are expanded

a distinct color shift occurs. The upper half of the coney's body flushes a darker brown, its belly fades to bright white, and its spots shrink to tiny black dots (Illustrations 115–118). If the intruder subsequently leaves, the coney soon returns to its normal coloration.

Even greater control of what is called cryptic or camouflage coloration (Illustrations 101–104) is available to the peacock flounder (*Bothus lunatus*). This remarkable flounder and many of its worldwide relatives can have sky-blue ocellations, fade completely white to lie on sunlit sand, or nubble itself brown to disappear into a background of coarse gravelly sand. In controlled environments, a flounder has even been known to adopt a checkerboard pattern when placed on a checkered surface.

The physical mechanism which makes such color control possible centers about the *chromatophore*, an irregular cell containing colored pigment. In several fish species, colored pigment can be shifted from the center of the cell to an outlying network of radiating processes, thus making that color more pronounced to the observer. This pigment shift is under both hormonal and neuromuscular control and is therefore an integral part of the animal's reactions of fear, aggression, stealth, or sexual arousal. Chromatophore pigments range through white, blue, green, red, brown, yellow, and orange, and their combinations make possible a complete spectrum of colors.

Groupers, goatfish, certain butterflyfish, and others are best categorized as having somewhat limited color control, two or three main color patterns constituting the entire repertoire. This reflects limitations in the number and kind of chromatophore cells as well as in the pigment-shift type of chromatophores. The flounder obviously has a greater variety of pigment colors in its pigment cells than the groupers or goatfish. This accounts for the flounder's ability to display such distinctively different colors as black, blue, brown, white, and yellow.

Another colorful family, the cephalopods (squid, cuttlefish, and octopus), makes the serranids and flounder look like amateurs. I was photographing one day off an isolated cay on the Great Barrier Reef when I noted out of the corner of my eye that a small coral head next to me was hovering a few centimeters off the bottom. I turned to look at it, and found it was staring right back at me with a large central eye. Since it was only a few centimeters away from me, I gingerly backed off and trained my camera on it. At the first flash of my strobe light one end of the "coral head" resolved itself into a mass of tentacles, while the other end began rippling a transparent fringe. I fired once more and the creature, now identifiable as a cuttlefish, began shifting colors, showing rich browns, blacks, yellows and, along its rippling mantle, an eerie iridescent blue (Illustrations 119–122).

These fantastic purveyors of color dynamics also use

the mechanism of chromatophores. The chromatophores of the cephalopods have two advantages over
those of the vertebrates: a functional design which
makes them more effective, and more colors—blue,
green, yellow, brown, and red—plus combinations of
these colors. When we add the reflecting qualities of
small cells called *iridocytes* scattered through the
epidermis, we get the iridescence which the cuttlefish, in particular, flushes when disturbed.

The chromatophores of the cephalopods are pigment-
filled bags scattered through the thin, translucent
epidermis of the animal, each with a set of radiating
muscle fibers. When these fibers contract under the
control of the central nervous system, the small bag
of pigment is stretched into a flat, highly visible disc
of that single color. Each disc may reach a few millimeters in diameter, though the sac when contracted
is so small as to be invisible. Some kind of phantas-
magoric emotional or mental state would seem to
control these pyrotechnic displays.

Outdoing even the photochromic sorcery of the cuttlefish is that of the octopus (Illustrations 123–124).
Not only does this animal possess the same wide-
ranging color control as the cuttlefish, but also it can
make its skin surface resemble the texture of coral
or coralline algae growths. When cornered, it can
find a tiny crevice and ooze its body through to safety.
Lack of a firm skeleton gives the octopus this ability.
The Cephalopoda also possess a considerable order of
intelligence. Captive octopuses have lifted the lids of
laboratory aquaria, crawled from their own to a
neighbor's, lifted the second lid, devoured the neighbor, and returned. This and other highly intelligent
behavior is due to the cephalopods' large, well-developed brain and nervous system.

Selective color filtration by sea water is operative at
all light levels except total darkness. In water deeper
than 15 meters the undersea world is generally perceived in monochromatic blue. The fantastic colors
we see in many underwater photographs are the
result of divers' lights, which restore the full spectrum of color at close range. Without such illumination, the photographs in this volume would largely
be seen in shades of blue and green. Yet those lavish
pigments are there in the undersea citizens when we
light them. There is much more to be explored concerning the colors in the sea.

101. *A peacock flounder* (Bothus lunatus) *adapting its body coloration to its white sand resting place. The fish frequently stirs up sand to cover part of its body, achieving almost total camouflage.* (*Cayman Islands, Caribbean*)

102–104. *The peacock flounder changing its color to match its background in an attempt to hide. It reverts to a pattern of pale blue spots when secure.* (*Cayman Islands and Bonaire, Caribbean*)

105. *The trumpetfish* (Aulostomus maculatus), *known for its use of color and body shape in hunting. Here the animal hovers amid the arms of a gorgonian sea whip. This fish is a swift predator despite its unprepossessing size and shape.* (*Curaçao, Caribbean*)

106. *The trumpetfish* (Aulostomus maculatus), *a species that comes in a variety of colors, including this golden yellow and brilliant blue.* (*Bonaire, Caribbean*)

103

104

106

107

108

109

110

111

112

113

114

107–108. Juvenile and adult coloration of the Caribbean blue tang (Acanthurus coeruleus). *(Both Curaçao, Caribbean)*

109–110. Juvenile and adult coloration of the Caribbean coney. (Cephalopholis fulva)

111–112. Adolescent and adult colors of the tiger grouper (Mycteroperca tigris). *A comparison of the juvenile with the adolescent of this species is found in illustrations 125–126. (Caribbean)*

113–114. Juvenile and adult coloration of the gray angelfish (Pomacanthus arcuatus). *The juvenile's stripes are consistent with its activities as a cleaner; the monotone adults do not act as cleaners. (Both Curaçao, Caribbean)*

115–118. The coney (Cephalopholis fulva) *changing from its normal to its alarm coloration. Its entire color repertoire can be seen in a matter of seconds. (Caribbean)*

119–122. The cuttlefish, a master of color control, displaying some of its repertoire. The cuttlefish has such quick control of its chromatophores (color cells) that it can literally flash waves of color across its body before one's eyes. (Australia)

123–124. The octopus, another master of coloration. The same individual is shown here in photos taken moments apart. (Rangiroa, French Polynesia)

125–126. Juvenile and adolescent coloration of the Caribbean tiger grouper (Mycteroperca tigris). *A comparison of the adolescent with the adult of this species is found in illustrations 111–112.*

127–128. The grouper Epinephelus fasciatus, *of the Red Sea, displaying two different body color patterns only moments apart. The color control of these fish seems akin to that of the Caribbean coneys in illustrations 115–118.*

15

119

16

120

17

121

118

122

123

124

125

126

127

128

After Darkness Falls

Most human forays into the underwater wilderness are conducted by day. That is our active period since we are dependent on the penetration of bright sunlight to illuminate the undersea scene. Much underwater photography is impossible after dark because there is no ambient light to observe action at any distance, and underwater lights are effective only over rather short ranges, due to light absorption by the water.

But what of the night, that other, forgotten half of the daily cycle? What happens in the underwater wilderness after the last sea birds have settled in their nests in the late afternoon?

The sun sets. Touching the tropic beaches, the calm of dusk leaves only gentle ripples to whisper along the sand. On the beach, human strollers take a moment to view the splendor, pausing in silhouette to contemplate the glowing dusk. Human activities have generally passed their peak and begin to taper off toward eventual sleep.

Under the darkening waters, where the coral reef is losing its last contact with the light, a great transformation is taking place, for the reef and many of its inhabitants are creatures of the night. We have already seen that the twilight periods, those dim corridors between daylight and darkness, are filled with hazard. During these brief hours there is on the reef a sense of tension that is almost palpable. Great stalkers roam the half-light, their obliterative countershading rendering them nearly invisible. Many species of open-water fish desert the water column during this "quiet period," moving nearer the reef structure for shelter. But the quiet period ends when the stalkers can no longer see accurately, and it is then that the nocturnal society comes awake.

One fundamental change that triggers this night-awakening on the reef is the emergence of the coral polyps. Although some feed during the day, most species wait until the darkness is complete. Then, armies of polyps numbering in the uncounted billions rise from their protective limestone fortresses and begin to sift the darkling sea for minute particles of food. Large stony coral heads become covered with delicate, soft tentacles (Illustration 130). The visual effect is a magical softening, almost blurring, of the hard look of corals as seen by day. Species such as the strawberry coral (*Tubastrea*) put on even more spectacular displays (Illustration 132). This coral resembles blunt brown fingers during the day, but in the night-dark water it extends brilliant orange bodies with tentacles that spread like flames.

Since the coral polyps themselves are food to a host of other reef inhabitants, a chain reaction of emergence quickly follows. Spider crabs, which hide deep within crevices in coral heads by day, march to the limestone battlements and begin plucking polyps with busy claws. Small, transparent shrimp with orange eyes that gleam in the diver's light move out to feed, as do the writhing brittle starfish (ophiuroids). For the brittle stars, daylight emergence would be fatal. Even when lured by food, these supple creatures will extend but a tentacle into the daylight from their coral sanctuary. If they are removed and set down in the open during daylight, they are instantly attacked by a horde of aggressive reef fish and devoured. During the night, however, the brittle stars swarm out into the open, festooning nearby sponges with hundreds of hairy feeding arms.

The sea urchin (*Diadema*) is another creature of the night. As the sun sets, these spiny organisms leave the crevices where they have hidden since dawn and move about all over the reef, browsing on algae. They even climb atop coral heads, menacing any fish that dare to move about. In the beam of our handlights their bodies pulse with brilliant reds and blues not seen during their daytime rest (Illustration 214). Atop bushy gorgonia or sponges the basket starfish begin to untangle. During the day, these fascinating animals form a Gordian knot of branched filaments, but in the darkness they spread a fan of arms often one-half meter or more in diameter, filigreed with hundreds of tiny protuberances forming an almost impassable barrier for their planktonic prey (Illustration 131).

Other invertebrates awaken. One corallimorpharian anemone spreads beautiful transparent tentacles, each with a bright red ball at its tip (Illustration 138). Waving these sticky, nematocyst-armed tentacles in the darkness, the anemone moves its central mouth in a circle so that the catch made by each tentacle is brought to it, in turn, to be consumed. Another anemone, *Cerianthus*, lies buried during the day. At darkness, this graceful animal rises within its tube, extending its mouth-topped body and a ring of sensuously waving long tentacles. In the shallow sand flats near shore, the *Cerianthus* wave gently back and forth in the subdued surge as if engaged in some ancient and endless dance (Illustration 137).

What of the vertebrates? What happens to the free-swimming animals of the reef? Amid the flowerlike gardens of night-feeding coral where are the hordes of chromis, wrasses, and other fish that throng the bright daylight waters?

Many of them are "asleep." As the light fades, one sees them all move closer to the reef and one by one find resting places in crevices and niches around the coral heads. Most of the smaller fish merely hide, but even at rest they exhibit some puzzling traits. Certain species of parrotfish exude a mucous envelope like some great balloon that stands out from the fish's body. Aquarium studies have shown that parrotfish trapped near a night predator, such as a moray eel, have vastly better survival possibilities when surrounded by their envelopes than they have without them. Scientists hypothesize that the mucous cocoon acts as an effective odor barrier to the scent-searching of nocturnal eel predators. Yet sometimes the parrotfish produce this mucous balloon and some-

times they do not. Many parrotfish wedge themselves against sponges as if blending their body contour against that of the sponge. I have seen others nestled into thickets of whip sponges so that only an eye showed in the glare of my handlight. I have even come upon a large rainbow parrotfish (*Scarus guacamaia*) sitting in the back seat of a sunken automobile on a deep reef off Curaçao.

Many residents of the daytime reef cannot be found even with intensive lighting of the coral heads, simply because they are not there. Most of the smaller wrasses, for example, which form one of the reef's largest families, bury themselves completely in the sand. Around the tropical world the inshore shallow sand flats are quite alive at night, though only a few species are visible to the diver.

Secretive fish such as the bigeye (*Priacanthus arenatus*), the glasseye (*P. cruentatus*), and the tiny cardinal fishes (Apogonidae), which lie hidden in the coral all day, move freely about in the darkness. Spiny pufferfish (*D. holocanthus*) emerge from their coral fastness and browse on the tiny shellfish of the coral shallows in apparent safety.

Among other nocturnal species, I once discovered a spiny starfish in a brilliant red color browsing on an algae-covered boulder off Fernandina Island in the Galápagos (Illustration 129). In other locales I have found surgeonfish, emperor snapper, and large queen triggerfish nestled into coral perches or propped against sponges—active hunters of the daytime reef, dormant in the darkness.

What of the predators? Is the nighttime reef a haven, a zone of truce in nature's endless hunt? Far from it. Morays and certain sharks are night hunters, and among the smaller denizens the hunt is as swift and deadly as in the day. Among the large reef predators, the one most likely to be encountered is the moray eel, sometimes swimming in the water above the reef, sometimes draped over the coral like some toothy noodle. Woe to the small fish that wanders past that noodle, however. This might explain why the small reef fish seem to quiver in the diver's light and never leave their coral crevices.

For a modern diver, night on the reef is an adventure in color and action. Armed with powerful lights, he finds the startling and the beautiful with every move he makes. The diver is more than an observer. The very presence of a diver on the reef catalyzes the action. Every creature reacts to the thundering noise of the diver's exhalations and the fierce glare of the lights. Man on the reef upsets a delicate balance, sometimes in favor of the hunter, sometimes of the hunted.

For example, one night I was diving with an underwater cinematographer, filming a barbershop shrimp beneath a small coral outcrop. The strong, hot lights attracted a swarm of fry—tiny fish not long out of the egg. In the middle of filming, an arrow crab marched into the scene. Normally this crab is a timid daddy longlegs. But not under the bright lights in the Hollywood atmosphere. With the camera grinding, the crab stomped to center stage and began snatching the fry out of the suddenly bright water with both blue-tipped feeding claws. One after another of the fry disappeared into its greedy maw, until the camera ran out of film and the lights went out. I have never felt the same about arrow crabs since.

Sometimes the effects are hair-raising. On one night dive in extremely clear water under a full moon, the reef was so bright that the divers turned out their lights spontaneously and swam along fascinated in the ghostly light. Suddenly a large shadow moving in the water led us to turn on the lights, revealing a reef shark startled at being spotted. With an instinct shared by many other animals, the shark came directly towards the nearest light and bumped it with its nose. The shark then raced off into the night in a flash of phosphorescence.

The instinct of creatures to "home in" on a light has yielded me other interesting night experiences. In one, I was 50 meters deep on a reef in Curaçao. Suddenly, my probing beam found a large shrimp eel, normally a retiring species. Instantly, the eel raced wriggling up the beam like some snake on a tree branch, ascending ominously toward me. The eel never stopped until it crashed head-on into the light.

Young turtles often seek nighttime shelter amid the branches of large, bushlike gorgonian colonies, apparently in hiding from "vibration sensors" such as sharks. One night I turned my beam on just such a sleeping youngster. It sought out the light several times, crashing into the bulb on each occasion. After each strike it would swim off into the darkness, only to return when impaled by the bright beam.

There is no question that a powerful light, properly used, is the key to successful underwater photography at night. I have often illuminated a fish with a bright handlight and watched it "freeze," possibly fearing the terrible spines of the omnipresent sea urchins, which can be injurious or even fatal. (Often I have seen fish or eels with a blinded eye and a characteristic riddling pattern on the side which bespeaks a desperate encounter with those spines.)

When lit, fish such as the butterflyfish and trumpetfish hold all their fins extended as feelers and move toward shelter very gingerly. A careful diver can use his free hand to guide the fish up into open water, where the urchins are also less of a hazard to the photographer.

Not all fish react to the light beam with caution. Some, such as snappers and parrotfish, will freeze for a few moments, then abruptly bolt, crashing repeatedly into coral heads, urchins, and even human intruders. I never try to precipitate such panic, lighting these species only briefly and not attempting to handle them.

One family of fish which offers unusually fine photographic opportunities at night are the porcupinefish, spiny puffers, and burrfish, which can be immobilized by the light, captured (whereupon they will water-inflate themselves in panic), and photographed without harm.

As mentioned in the preceding chapter, color and color change are a widespread and important feature of the nocturnal society of the reef. Many fish adopt what would appear to be some variation of disruptive coloration when sleeping. The creolefish (*Paranthias furcifer*), for example, has a uniform gray-roan body color during the day. At night, however, it displays spots and stripes of white as it nestles in comfortable sponges or corals (Illustrations 156, 158). Similarly, the red hind (*Epinephelus guttatus*), one of the groupers, subdues its spotted colors and flashes broad vertical stripes. Squid can be found in brilliant scarlet, octopus in a rainbow of changing colors, and even the dull goatfish in red, pink, and iridescent blue (Illustrations 149, 151).

There has been much disagreement about the different colors that certain species of fish display at night. One school holds that the intrusion of divers with powerful lights causes the fish to exhibit alarm or stress colors. This might well be true in such fishes as goatfish. Another theory is that night-feeding species may be displaying feeding colors rather than nocturnal hues. Among certain grunts and other evening browsers that, too, may be true. A third theory is that the light beam itself reflects back different colors than does the daylight sun. Again, there are some types of light sources and some species of reef life for which this seems to be the case.

A wide range of species seems to fit none of these categories. The masked butterflyfish (*Chaetodon lunula*), for instance, swims about in splashy finery of canary yellow and black, but in the darkness the yellow is completely drained to dull white (Illustrations 154–155). Then there is the spotfin butterflyfish (*C. ocellatus*). By day its body is a yellow-rimmed white disc with a tiny black dorsal spot and a black eye-stripe. In the darkness, however, it is adorned with a large eyespot on its upper back and large dark swaths diagonally across its body. Neither of these fish feeds by night, and I was not able to induce color change, either, by putting them under stress or under lights by day. Their nocturnal coloration appears to be just that, a totally altered raiment for darkness. The same can be said for a number of other species, such as surgeonfish, tangs, parrotfishes, bonnetmouths, and triggerfishes.

Lobsters, one of the most favored delicacies of the sea, are abroad during the night. The spiny and slipper lobsters can be found under gorgonians and coral heads even in areas where they are never seen during the day. Natives of many islands walk along the shore with lanterns, pouncing on the secretive Indian lobsters revealed in the flickering light.

The ultimate sense of night on the reef is reserved for

those divers who extinguish their lights. In the darkness, the slightest movement triggers a swish of phosphorescence which seems only to deepen the blackness. The noises of the diver's bubbles are fearfully loud as each imagination goes its own way. Soon the lights begin to come on and the swim is resumed. The memory fades slowly, linked as it must be to our ancient fear of what prowls in the dark void.

For all the beauty of the evening reef, there is an unavoidable evocation in the human observer of violence and sudden death. Consider the graceful *Cerianthus*, the burrowing anemone of the sandy shallows. A diver who takes a bright light into the shallows is often surrounded by swarms of tiny wrigglers, open-water worms sensitive to light. They flash by on all sides, swimming in such panic that they will crash into the diver and even wriggle into the collar of a rubber suit. If the diver beams the light on a *Cerianthus* while these tiny creatures are roiling frantically about, the streaking wrigglers swim by the hundreds right into the waiting tentacles of the anemone. In a moment the sensuous dancer with the long fingers becomes a predator, as one tentacle after another wraps around a worm. In a few seconds the *Cerianthus* is a wriggling ball, receding into its tube to consume its unexpected meal.

This is the drama of the night on the tropical reef: an intoxicating blend of mystery, brilliant color, comedy, and suppressed violence; when "things go bump in the night," and you imagine something large and dark over your shoulder or around the next bend of the reef; when colors change, and the fish you saw that afternoon looks totally different; when hobgoblins loom with arms outspread, only to become basket starfish with tentacles that fold slowly in your light; when the vast ocean is only as large as the aura of your lamp. It is no time for the fainthearted or the armchair explorer, for at the very extreme you are alone with yourself.

129. *A starfish, at night, feeding on algae and crustaceans on a lava boulder.* (*Galápagos Islands*)

130. *Polyp tentacles of the Caribbean pillar coral* (Meandrina meandrites), *commonly found in large colonies.* (*Roatán, Republic of Honduras*)

131. *The Caribbean basket starfish* (Astroglymma sculptum). *It forms a tangled, hidden mass by day. At night it mounts a coral perch and extends its intricate arms across as much as a square meter of food-bearing current.*

132. *Night-blooming polyps of the coral* Tubastrea tenuilamellosa. *This species is very abundant in the Netherlands Antilles but uncommon in other Caribbean regions.* (*Curaçao*)

133. *The Caribbean coral* Eusmilia fastigiata, *with tentacles extended for night-feeding. The entire polyp is no more than two centimeters across the widest part of its body.*

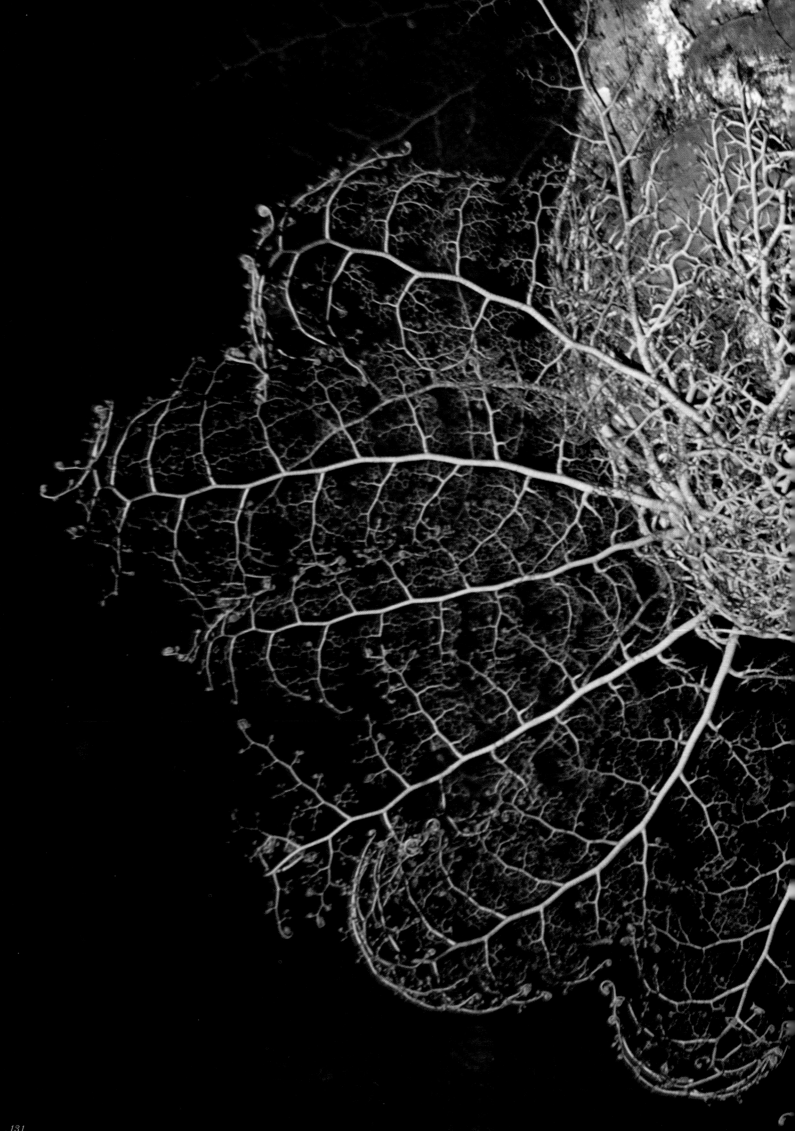

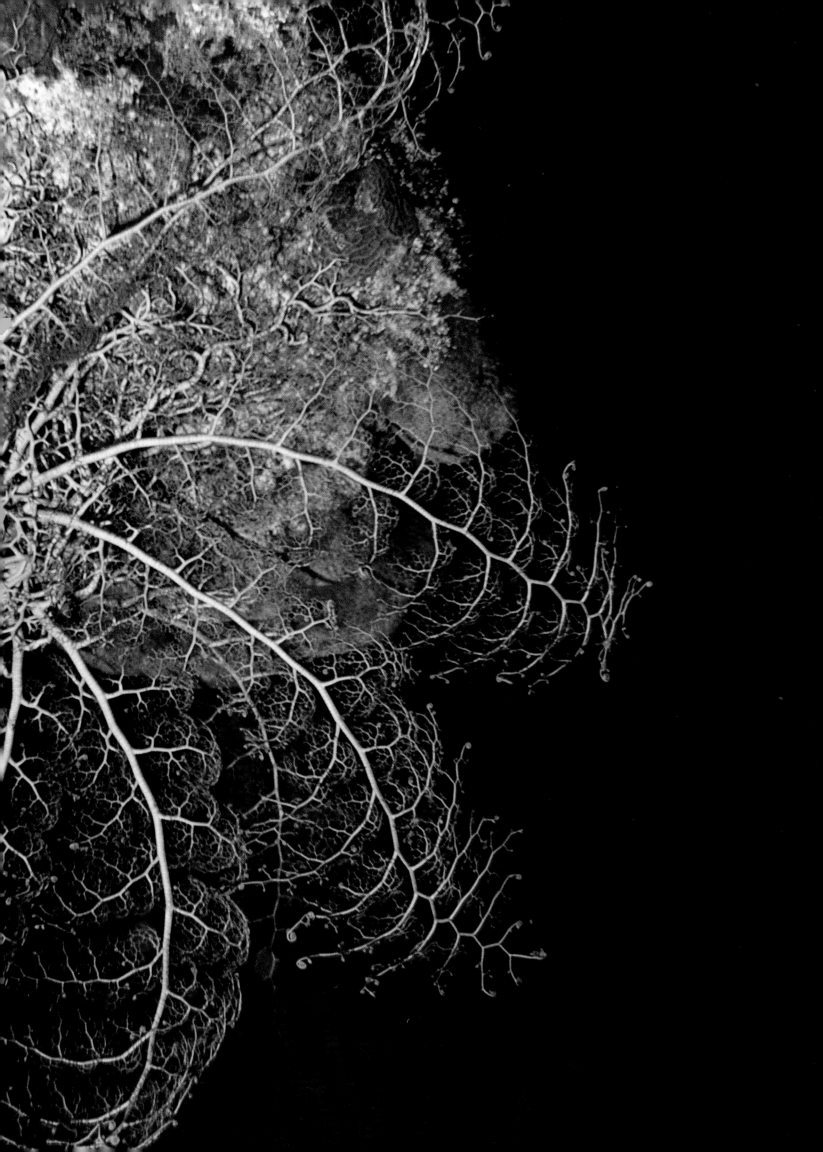

134. Polyps of the coral Dendrophyllia. *(Red Sea)*
135. The basket starfish (Astroglymma sculptum), *an intricately-fashioned creature, which opens only in darkness. Basket starfish may reach one meter in breadth, a startling sight on a dark reef. (Cozumel, Mexico)*
136. The feeding arm of the brittle starfish Ophiotrix, *another nocturnal feeder. (Bonaire, Caribbean)*
137. A cerianthid tube anemone rising above the Caribbean sand at night. It quickly disappears when approached. Even the slightest illumination from a diver's handlight causes the anemone to withdraw. (Bonaire, Caribbean)
138. This species of nocturnal corallimorpharian anemone was discovered by scientists within the past decade. (Curaçao, Caribbean)

135

136

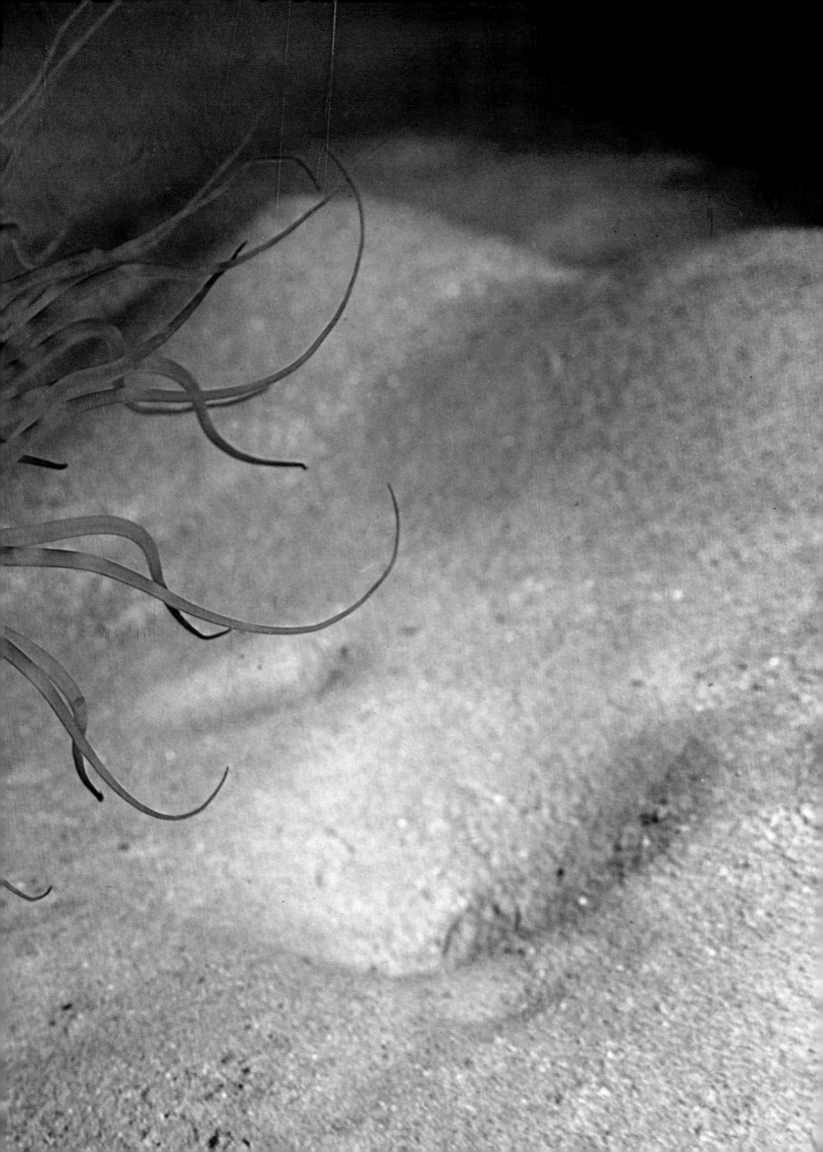

to the sea floor. (*Cayman Islands, Caribbean*)

141. *The spotfin butterflyfish* (Chaetodon ocellatus), *in its nocturnal coloration. Its daytime colors are shown in illustration 25.* (*Cayman Islands, Caribbean*)

142. *The squirrelfish* (Holocentrus rufus), *a night-feeding species which hovers in shadow by day. Its large eyes are well adapted for night vision.* (*Cayman Islands, Caribbean*)

143. *The queen triggerfish* (Balistes vetula) *in its night colors.* (*Cayman Islands, Caribbean*)

144. *A Caribbean scrawled cowfish* (Acanthostracion quadricornis) *in night colors.* (*Curaçao, Caribbean*)

145. *Red emperor sweetlip* (Lethrinus), *bedded down on an Australian reef at night.*

146. *The Hawaiian surgeonfish.* (Naso lituratus). *The spotted pattern is visible only at night.* (*Kona Coast, Hawaii*)

147. *The web burrfish* (Chilomycterus antillarum) *caught in lights as it browses on a dark Caribbean reef.*

148. *Sleeping hawksbill turtle* (Eretmochelys imbricata), *photographed at dusk. It retained its pose despite the great disturbance made by the divers.* (*Roatán, Republic of Honduras*)

149. *A goatfish* Parupeneus, *resplendent in its night colors, nestling in its coral refuge. The same species is seen by day in illustration 151.* (*Hawaii*)

150. *The glasseye* (Priacanthus cruentatus) *in its daytime colors. Its nocturnal coloration is similar to that of the bigeye in illustration 152.* (*Curaçao, Caribbean*)

151. *A goatfish* Parupeneus, *shown in its normal daytime colors. Its nocturnal coloration is seen in illustration 149.* (*Hawaii*)

152. *The bigeye snapper* (Priacanthus arenatus), *a purely nocturnal fish, that generally hides in crevices during the day. Its daytime colors are similar to those of the glasseye in illustration 150.* (*Cayman Islands, Caribbean*)

153. *A parrotfish nestling under the spines of the urchin,* Diadema setosom. *The urchin uses its dangerous spines to defend itself.* (*Cayman Islands, Caribbean*)

154. *The masked butterflyfish* (Chaetodon lunula), *in its normal daytime coloration. Its strikingly different nocturnal colors are shown in illustration 155.* (*Hawaii*)

155. Chaetodon lunula, *the masked butterflfish, in its nocturnal colors, a sharp contrast to its daytime coloration shown in illustration 154.* (*Hawaii*)

156. *The creolefish* Paranthias furcifer *as seen by day. A view of the very different nocturnal coloration of this species is shown in illustration 158.* (*Bonaire, in illustration 156.*

158. Paranthias furcifer, *the creolefish, in its patterned night colors, resting in a sponge. This species in its daytime colors is shown in illustration 156.* (*Bonaire, Caribbean*)

159. *A stoplight parrotfish,* Sparisoma viride, *in its night colors, resting.* (*Bonaire, Caribbean*)

141

142

143

144

145

146

147

148

149

150

151

152

154

153

155

156

157

158

Part Two

The World's Greatest Underwater Wilderness Areas

A Garden of Islands: The Caribbean

Having considered some aspects of the life and historical background common to all underwater wilderness areas around the world, we now turn to one of the most beautiful of these marine provinces, the Caribbean. Its proximity to the Americas, its wonderful climate, and its tropical grandeur have made this sea irresistible to an ever-growing number of divers, sailing enthusiasts, and vacationers.

Many of the same species of corals, gorgonians, fish, turtles, and so forth are found throughout the Caribbean, but there are subtle differences in abundance from island to island. One island or zone may appear exceptionally rich in sponges, another in gorgonians, another in pillar coral, and still another in blue chromis fish. I will attempt to highlight the most distinctive species in each zone of this fascinating sea.

The first important discovery we make concerning this tropical western Atlantic province is that its fauna is uniquely different from that of other seas. The explanation of this phenomenon lies in two circumstances in the geologic history of the region. First, we now know that Central America has long been an arena of severe tectonic activity. This was evidenced again as recently as 1976 in the fearful earthquake in Guatemala. In past ages, even more violent episodes have resulted in the repeated engulfing and emergence of the Isthmus of Panama, the land bridge between North and South America. As a result, the Caribbean marine fauna has alternately undergone periods of intermixing with species of the eastern Pacific, and periods totally devoid of contact between them. Currently the Caribbean is in a period of isolation which has lasted well over a million years.

A second major effect occurred during the glacial epochs. The Caribbean, more than any other region, demonstrates what is known in geology as the glacial control theory. The reefs here tend to be built on areas leveled by wave erosion during eras of glaciation. Some of these reefs, such as those of the Bahamas, are many kilometers in width. There are very few atolls or barrier reefs in the Caribbean; those are more characteristic of subsidence zones such as the South Pacific.

Everywhere in the Caribbean one sees the effects of the different sea levels during glacial and interglacial periods. Islands ranging from those in the northernmost Caribbean to those off the northern coast of South America display the characteristic "lunar" coastlines of jagged, eroded limestone. These are the exposed coral skeletons of older reefs which grew during the higher sea levels of an earlier period. As for the living Caribbean reefs, most estimates place their age at only 5,000 to 10,000 years, or since the last glacial era. At that time, sea levels were far lower than they are today, due to the incalculable volume of water taken up in the vast ice sheets covering the hemisphere as far south as the state of Pennsylvania. During that glacial epoch, all present-day

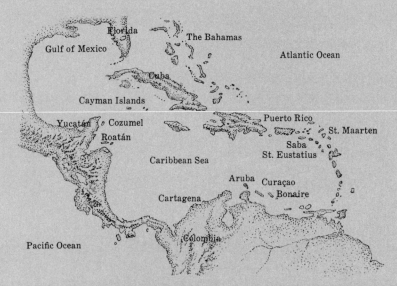

Among the islands of the Caribbean the reefs range from vast expanses of coral-filled shallows to coral-lined undersea precipices that plunge thousands of meters to the sea floor.

Caribbean reefs were high and dry—30 meters or more above the sea level at that time—and subject to tremendous erosion.

What we therefore see in the Caribbean underwater wilderness are young coral reefs dominated by species which rose to prominence over a short span of evolutionary time and in an isolated, semi-enclosed sea in conditions very much like the present. Thus, despite the prolific life on Caribbean reefs, scientists consider the fauna impoverished as compared to the older, larger, and richer Pacific fauna. For example, the Caribbean is known to have 27 genera and 48 species of corals; the Pacific boasts 90 genera and some 600 species. Among other families, scientists have reported 92 separate species of damselfish in Australian waters; the Caribbean has only 12.

The Caribbean's geological history also accounts for the prominence of certain species among its fauna, such as gorgonian corals, the familiar sea whips, and sea fans—species which are less significant in other seas. Certain species will be numerous on certain islands due to local conditions. The Cayman Islands, for example, show incredible abundance in both the quality and variety of sponges on their reefs. In Bonaire in the southern Caribbean, and in the Bahamas and Florida Keys in the northern Caribbean, conditions are right for extraordinary growth in species of both *Acropora*, the branching stony corals, and gorgonians.

To illustrate the variety of Caribbean reefs and reef inhabitants, I shall focus on the following areas: Cozumel Island and other reefs of the Yucatán and Belize coasts; Roatán Island, Republic of Honduras; Cartagena, Colombia; Curaçao and Bonaire, Netherlands Antilles; Swan Island, central Caribbean; Cayman Islands; the Saba Banks near St. Maarten; and Nassau, Andros, and other islands of the Bahamas.

Cozumel and the Mexican-Yucatán Coastline Southward to Belize

When scuba-diving began in earnest in the Caribbean during the early 1960s, there was one "ultimate reef," the legendary Palancar Reef off the southwest coast of Cozumel Island. Cozumel is a flat limestone island 30 kilometers long and some 20 kilometers off the coast of the Mexican Yucatán. A thousand years ago, Cozumel was a sacred site for the Mayan Indians. Modern exploration has finally uncovered the full extent of the Mayan civilization. As the ruins of enormous cities such as Uxmal, Chichén Itzá, and Coba are investigated, estimates of the Mayan population have risen to as high as 20 million. And yet no one knows what happened to these tremendously sophisticated builders, traders, mathematicians, astronomers, and priests. The Mayan civilization and its huge population simply disappeared, and there remain only awesome temples towering above the jungles of Central America.

Cozumel Island was the focus of another major event in the history of the hemisphere. It was here that

140

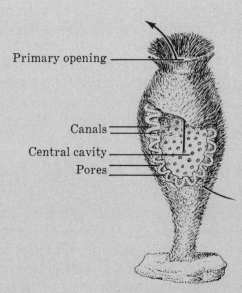

Cross section of a sponge, the most primitive of multicellular animals.

Primary opening

Canals

Central cavity

Pores

Hernando Cortez assembled his few hundred conquistadores for the remarkable march to what is now Mexico City and the subjugation of the entire nation. Today the palm trees of Cozumel's sun-drenched coastline are slowly giving way to beachfront construction. The island has become a bustling tourist center peppered with hotels and condominiums. Large cruise ships now make it a regular port of call. About a kilometer off the southern end of Cozumel is one of the natural wonders of the undersea world, Palancar Reef. Here on the boundary of the island's offshore shelf, overlooking the deep water of the Yucatán Channel, has grown a massive fortress of coral. The shelf is essentially linear, dropping steeply from its shallowest level of 10 meters to a depth of 400 meters. Along the edge of this drop-off is an epic bastion of coral, some five kilometers long, punctuated by huge canyons, overhangs, and caverns. On the slope of this dizzying precipice grow towers of stony coral, free-standing 30 meters tall or more in the open water.

The first time I swam out over this abyss in 1967, it was as breathtaking as flying out over the Grand Canyon. Since that day, thousands of divers have experienced the thrill of this famous reef. In a very real sense, Palancar Reef today is a vast sepulcher because, during the 1960s, ardent and unthinking spearfishermen slaughtered the larger reef fish for sport. They were followed by a wave of collectors who tore out many thousands of black coral "trees," used to make coral jewelry. Today, a diver must descend more than 70 meters to find even a small colony of this precious coral.

The story of black coral is another ecological tragedy. Although very little of each tree is actually suitable for jewelry, collectors destroyed entire colonies to get a few select branches. Later collectors ripped out the trees just to prove they were deep divers, so that many colonies perished for nothing. No one realized until too late that the coral could have been harvested, yielding all the branches wanted for jewelry and yet continuing to grow new stock. It is too late now.

This first wave of destruction took place when Cozumel was a remote wilderness. It preceded the realization that even such massive reefs as Palancar are highly vulnerable to the effects of human predation. Since that realization, the depredations have diminished somewhat. This is due partly to an increased understanding of ecological problems and partly to the greatly reduced population of both gamefish and black coral.

There has been another, later wave of carnage, one that is inflicted unintentionally by thousands of innocent divers, including many conservationists. Inexperienced divers inadvertently maul the coral with their hands, elbows, or knees; others kick sand (which cannot be safely ingested by the coral) over it with their swimfins; still others rip off "just a little

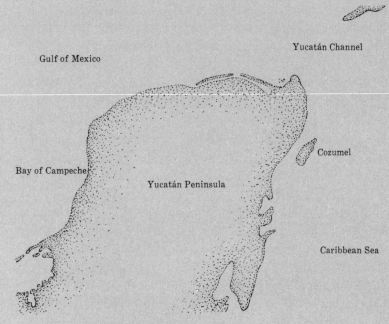

Cozumel and the Mexican-Yucatán coastline

piece" to adorn the coffee table back home. Thus a reef that becomes popular will inevitably suffer damage even at the hands of its admirers.

Palancar Reef is still awesome, still heroic. There is still hardly a thrill in the underwater world to compare with moving from cavern to cavern in the darkness, realizing that you are moving entirely within the great reef's structure. The entire five-kilometer labyrinth is honeycombed with tunnels and chambers, with small side passages leading to brilliant sunlit sand flats on the landward side, and to a vast electric-blue void on the ocean side.

While Palancar has retained its structural magnificence, its grandeur today is like that of the Parthenon, a great ruined temple, a reminder of splendors forever lost. For the new diver, its size is a superb example of the gigantic scale of building achieved by coral polyps, and it is still heart-stopping in its silent immensity. But for the diver who experienced this reef a mere decade ago, Palancar is a warning. If this could happen to Palancar, it could happen on any island, in any sea. Palancar is a part of the underwater wilderness which was not isolated enough, whose peril was not perceived soon enough, whose nation did not early enough see fit to protect it. Moving southward along the Yucatán coast, the next offshore reef structure lies nearly 160 kilometers away. This is the deserted Banco Chinchorro, or Chinchorro Banks, refuge of fishermen and latter-day pirates. Its deserted, lonely, birdswept beaches have looked out on the shipwrecks of Spanish galleons and modern freighters. The reefs of Cozumel, of Chinchorro, and Belize (Glover's and Lighthouse reefs) all share essentially the same fauna. Only the immense and complex turrets of Palancar are unusual. If I were to select a noteworthy family of marine life which is predominant in this Yucatán region it would be the building-dome corals.

Chinchorro, Lighthouse, and Glover's reefs are well offshore and fairly distant (at least 80 kilometers) from human population centers. For this reason, both their coral and resident fish populations are in better condition today than those of Palancar Reef. At these remote southern reefs we see in profusion the fish species we formerly saw at Palancar: horse-eyed jacks, sweeping schools of anchovetta, pairs of queen angelfish, French and gray angelfish, solitary barracuda, large tiger groupers, Nassau groupers, coneys, hamlets, tangs, rainbow and stoplight parrotfish, four-eyed and banded butterflyfish, squirrelfish, blue chromis, eagle rays, and a host of others. We will meet these species again and again in the coming pages, for they are among the most widely represented species on all Caribbean reefs.

The coral structures along these characteristically western Caribbean reefs are composed of low, dense, strong colonies of dome corals such as *Montastrea, Porites, Siderastrea,* and *Solenastrea.* Surmounting these in areas of shallow water are the intricate

142

staghorn coral (*Acropora cervicornis*) and, in windward turbulent areas, the broad fronds of elkhorn coral (*A. palmata*).

A unique feature of Lighthouse Reef is the enormous "Blue Hole." In the center of an expanse of shallows that are punctuated with dome corals is a sudden, breath-taking cobalt-blue "hole" some 70 meters or more in diameter. Divers have penetrated the hole to depths of more than 70 meters, reaching a series of galleries with stalactites and stalagmites, remnants of an era when these caves were above sea level. Depth sounder readings exceeded 120 meters in the center of this remarkable, isolated phenomenon.

The Island of Roatán, off the Coast of Honduras

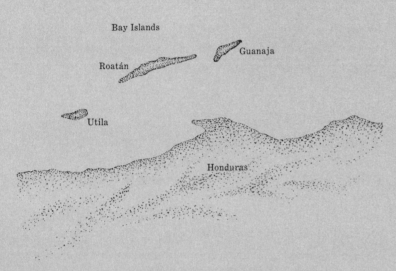

As we move southward and eastward from the Yucatán's offshore reefs we come to a group of scenically beautiful islands which have exquisite reefs. These are the Islas de la Bahia, or Bay Islands, of Honduras. The main islands are Utila, Guanaja, and Roatán, with the latter the best known to underwater enthusiasts. This long, rather narrow island lies pointed into fairly consistent winds coming from the east. As a result, much of Roatán's eastern coastline suffers from roughness and murky water conditions, though the corals are well developed. The finest of the island's reef areas available with good diving conditions are those of the northwestern coast. They are truly superb.

There is a noticeable difference in certain species inhabiting these reefs compared with those of the Yucatán, a mere 160 kilometers to the north. For example, Roatán has stands of pillar coral (*Dendrogyra cylindrus*) which are two to three times the size of colonies on other Caribbean reefs (Illustration 164). Some of the pillars, three to five meters tall, dwarf human visitors. Not only are these colonies extremely well developed, they also are scattered in great abundance across the reef in depths of 5 to 20 meters.

The blackcap basslet (*Gramma melacara*) is another abundant inhabitant of Roatán found only on the deeper reef slopes (Illustration 165). It is a purple and black relative of the royal gramma we met in the first chapter. In Roatán large numbers of this species are to be found at depths of 60 to 70 meters at the edge of the island's most spectacular drop-off, located along a six-kilometer expanse beginning at the island's western tip. Along this coastline are dive sites which have not yet become accessible to large numbers of divers; for this reason they are still in good ecological balance. The reef structures here are not nearly as immense as those of Cozumel. In a few spots, however, some of the large formations have developed tunnels, canyons, and caverns.

A warm memory for me is of a dive at dusk at Westend Bay when I discovered a young hawkbill turtle. The turtle lay completely motionless during 15 minutes of agitated strobe-light photography, doggedly refusing to flee into the surrounding darkness. As my companion and I returned to our boat we had to admire the animal's stoicism and endurance. We could only imagine what our reactions would have been in its place.

Cartagena, on the Coast of Colombia

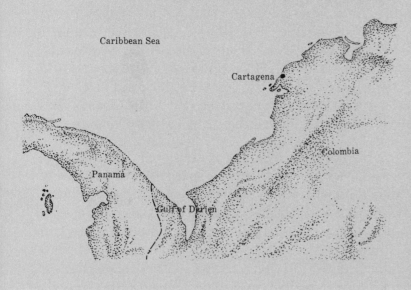

Continuing southward, we come upon the reefs of Cartagena, an example of reef conditions near the mainland coasts of the Caribbean. Cartagena's reefs are subject to enormous flows of silt-laden water from the mountains of Colombia's interior. At times this causes a shallow, muddy freshwater layer to spread across the surface of the sea, resulting in very reduced light penetration to the corals of these areas. For this and other reasons the predominating corals are extremely thin and fragile in structure. A diver can thrust his hand into this coral thicket up to his elbow, breaking off fine, brittle branches. The fragility of this coral, combined with the local practice of using dynamite in fishing (there are many one-armed cigarette-sellers in the city who held the dynamite too long), explains hundreds of bowl-shaped blast areas that pock the delicate reef.

The crinoids are another notably abundant family in this region. An unusual orange-brown color variant of the crinoid *Nemaster rubiginosa* contrasts vividly with the black and white version that predominates elsewhere. Other lacy crinoids are found here in great numbers and in a variety of colors.

The Southern Netherlands Antilles: Aruba, Curaçao, and Bonaire

These islands were my home for three years, during which I spent part of every day underwater. Having come to know their fauna, ecological conditions, and political history extremely well, I can use this knowledge to interpret the reefs of many other Caribbean islands.

First, a word about Caribbean history. Since the earliest explorations in the late 1400s, these islands and others like them have endured passing conquerors, including Christopher Columbus, Alonso de Ojeda, Vasco Nuñez de Balboa, Francisco Pizzaro, and Hernando Cortez. The major civilizations of the hemisphere, Inca and Aztec, were mercilessly overthrown and looted, their art and culture eradicated in the name of Church and King. The Antilles were hard hit. Their Indian population, the peaceful Arawaks, died in the Spanish mines of Hispaniola and were eventually replaced by African slaves, the basic stock of today's population. Cities such as Cartagena became infamous for the plunder which coursed through them on its way back to Europe. Of course, the Caribbean exacted some toll of the looters. The Florida shoals broke up dozens, perhaps hundreds, of their wooden ships and snuffed out the lives of their crews.

The conquerors destroyed not only the human inhabitants and cultures of these islands, but much of the wildlife and vegetation as well. Bonaire, known in the 17th and 18th centuries as Isla de la Palo Brazil, or Spicewood Island, by the 19th century had been degraded to Goat Island. The reason is that waves of conquerors—Portuguese, Spanish, English—had released hordes of the voracious animals to breed and serve as food for their military garrisons on Curaçao. The spicewood and other vegetation was destroyed by the marauding goats. Today cactus is the principal greenery of Bonaire.

The reef communities surrounding many Caribbean islands have fared little better. Growing native populations turn to the reefs for both food and sport. Shallow-water fishes are taken by primitive traps or nets. Deeper-dwelling species succumb to baited hand lines. In recent years, the advent of spearfishing and the use of oceanic lines sometimes kilometers long have put tremendous pressures on certain species. On some industrialized islands such as Curaçao (with a major oil refinery), pollution has wiped out or seriously damaged many kilometers of reef community.

Yet the tremendous resiliency of reefs is such that slipping beneath the waters off islands such as Bonaire is still a startlingly beautiful experience. This is one of the richest reef communities in the entire Caribbean. Vast stands of elkhorn and staghorn coral bloom in the sun-filled shallows, while swarms of reef fish, including chromis, black durgons, parrotfish, small groupers, trumpetfish, trunkfish, and others, flash and sparkle among the intricate protective maze of corals.

The undersea world of the Antilles is rich and diverse. For example, among the waving arms of the plentiful shallow-water gorgonians we find a small, leaf-shaped, well-camouflaged fish known as the slender filefish (*Monocanthus tuckeri*). This shy little creature carefully maneuvers among the branches of gorgonian sea whips to conceal itself as the photographer approaches, yielding its portrait only after an extended contest.

One particularly successful group in the Antilles is that of the gorgonians (Illustration 168). Not only do these tall, graceful sea whips fill the shallows, their relatives also bloom in unimaginably bright colors on sand-coated shelves more than 70 meters deep. These deep-water colonies differ from their shallow-water relatives in several important ways. Their polyps are larger and are scattered more sparsely along their skeletal arms than those of shallow-water gorgonians. Most impressive to the human observer is the amazing color in the fleshy tissue which covers the skeletons of these colonies. In the eternal twilight of the deep reef, the colors seem without purpose. Why here? Why in this Stygian darkness where red is merely black and yellow is light? These gorgonians are not the only unusual inhabitants of the deeper sand shelves. One day I was diving in a sandy valley off the coast of Bonaire. While brushing away the sand near an interesting shelf, I was suddenly confronted with a small sand cave-in. Moving in closer to inspect the five-centimeter bur-

Aruba Curaçao Bonaire

Gulf of Venezuela

Venezuela

Lake Maracaibo

row, I found myself face to face with an astonishing creature. This obviously agitated reef-dweller resembled a large white praying mantis with eyes like capsules on stalks. These capsular eyes seemed to be filled with moving black and white specks, and they rotated about the axis of their supporting stalks. When offered a morsel of bait, the tiny predator snapped it up. I later learned this was the fierce mantis shrimp (*Squilla*), and that its pincers—in fact a hinged lower jaw—had earned it its colloquial name, "Thumbsplitter."

Another remarkable denizen of the Antillean reefs is the longlure frogfish (*Antennarius multiocellatus*). These strikingly structured predators (Illustration 172) entice their prey within range by waving a slender lure mounted above their mouths. One day we encountered not one but two large, brilliantly yellow specimens. Moving them together for photographs, I suddenly discovered water leaking into my camera housing. I returned next day with an alternate camera and luckily encountered the same frogfish once again. Excitedly I began blazing away, only to see water droplets once again within this camera's housing. It is hard to believe, but I returned to that area a total of seven days and found the frogfish all seven times. Each day the housing leaked. All my efforts to repair and test it were futile, and the taciturn frogfish endured with what seemed like endless patience. Of course the day the frogfish disappeared, I discovered the problem and repaired the housing. I never saw the frogfish again.

During another fascinating encounter, I intervened in nature's normal process slightly by placing a hungry triton trumpet on the same coral head as an appetizing short-spined sea urchin. The triton reared up as if to look for the prey he sensed. Soon after, he started oozing across the coral until he reached the urchin, which had suddenly begun to move away. At the first touch of the urchin's spines, the leading edge of the triton's foot formed itself into a concave shape, grasped the urchin's spines by the tips, and lifted the helpless prey clear of the coral. Thus held aloft, the urchin could no longer escape. Slowly the triton rotated the urchin until it located the vulnerable mouth on the underside of its prey, and immediately probed through the mouth with a suddenly extended proboscis (Illustration 174). Enveloping its prey, the triton rolled off the coral onto the sand below and proceeded to extract the urchin's meaty interior. Soon, all that was left of the urchin was skin, spines, and jaws.

Swan Island

Swan Island actually consists of two islands lying between Grand Cayman in the north and Honduras in the south. Swan is surrounded by open ocean thousands of meters deep. It is also the site of a National Oceanic and Atmospheric Administration (NOAA) weather station which broadcasts daily weather reports to ships throughout the Caribbean. While its

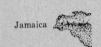
Jamaica

Swan Island

Caribbean Sea

Honduras

Nicaragua

Swan Island

recent history is tame (only a small NOAA staff and some Colombian workers live there now), its reputation in the 1960s was rather lurid. There are persistent tales of Swan as a CIA facility broadcasting into South and Central America. For a diver, such history seems remote anyway.

Through the calm surface of the waters surrounding Swan, we see enormous coral heads sitting on broad ridges, separated by narrow, sandy valleys. Entering the water, we find familiar species of shallow-water stony corals, but they are two or three times their normal size. One interesting coral that occurs in low-growing, intricate beds is a jagged hybrid that seems to be a cross between the common staghorn and elkhorn corals found elsewhere in the Caribbean.

The outstanding impression received from these islands is one of great vitality and good health, with their vibrant, plentiful fish populations, a goodly number of larger predators, and unbroken, well-developed corals.

Moving north and east from Swan, the next scene of our underseas travels is the Cayman Islands. According to recent investigations of plate tectonics, Grand Cayman, the largest of the three islands, is located at the edge of a tectonic plate. Directly off the southern coast of the island is the Cayman Trench, plunging eight kilometers to the eerie abyss where the American and Caribbean plates meet. There is no hint on the sun-dappled shore or even at the drop-off some kilometers offshore that we are diving at the edge of an escarpment which eventually plunges 5,000 meters to a highly active fault zone.

The Caymans are bustling islands. Their people are taciturn and shy, speaking in a curious, lilting, clipped English. Caymanians, like Antilleans, have been seamen for generations. Formerly, much of the male population plied the seas while the island's sleepy life droned on. All of that seems over now. The Cayman legislature never passed laws to control the influx either of financial corporations nominally located on the islands or of numerous branches of international banks which concentrated there. Thus these islands, like others such as Curaçao, became tax havens. Sleepy Georgetown now boasts 40 banks, as well as a variety of corporations expressly created to keep tax money from the United States and other nations. The resulting boom in construction has created many jobs for the Caymanians and forever altered the tropical paradise. Whether this will endanger the lush marine life of the offshore reefs remains to be seen.

The marine life of Grand Cayman and its satellite islands, Little Cayman and Cayman Brac, invites superlatives. The most impressive undersea family here is that of the sponges. In these islands there are more sponges of varied sizes and colors than I have seen anywhere else in the world (Illustrations 178–179). There are red, orange, purple, green, brown,

Cayman Islands

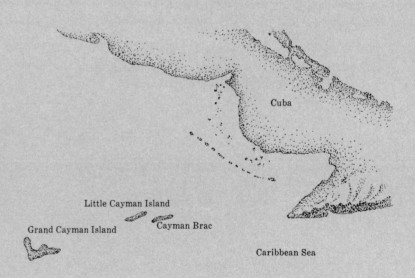

Cuba

Little Cayman Island

Grand Cayman Island   Cayman Brac

Caribbean Sea

and iridescent sponges. There are basket sponges large enough to hold one or even two divers. At depths of 70 meters off the plunging Wall of Little Cayman, there are orange sponges shaped like elephant ears that are more than two meters in diameter. At night, many of these sponges are festooned with the snaky arms of brittle starfish (ophiuroids). The sponges of the Caymans could comprise an entire chapter of marine life in themselves. The steep-sided undersea mountains that culminate in these islands have proved an ideal breeding ground for them.

The Wall at Little Cayman, mind-stopping in its immensity, is another phenomenon in a class by itself. The top of the drop-off begins a mere 10 meters beneath the surface. In describing the Palancar Reef drop-off, I mentioned the illusion of flying out over the Grand Canyon, with enormous coral towers, battlements, and canyons on every side. There is none of that here. This drop-off is like the edge of a table. As you swim out over it you see only the endless, electric-blue deeps. At about 45 meters down it even slopes inward as an undercut. When you hover near the huge sponges at 70 meters, your rising bubbles hit the reef face above you and cascade upward through the lush growths of coral, sponge, and gorgonians, and through schools of small fish. Of the more accessible drop-offs for divers in the Caribbean, the Wall at Little Cayman is perhaps the most spectacular. Other coasts in these islands, such as the north or south walls of Grand Cayman, are also quite steep. To reach them, the diver must descend 15 to 25 meters to the reef-top. This may require decompression after longer dives, a procedure that amateur or sport divers should shun if possible. The Little Cayman Wall with its very shallow crest avoids this problem, enhancing its attractiveness to sport divers.

One of Grand Cayman's primary attractions is to be found in a series of picturesque shallows at the southwest corner of the island. Here live a school of majestic silver tarpon, which hover in a valley filled with waving gorgonians. When divers invade the valley, these gleaming, wide-eyed hunters respond by slowly, placidly swimming to the next valley. If still followed by the divers, these streamlined fish move off once more. Confident in their speed and alertness, the tarpon lead the procession until the pursuing divers are exhausted.

The Saba Banks, Near St. Maarten

Far to the east of Cayman, at the northeastern corner of the immense island chain that extends from Cuba to Venezuela, lie the northern Netherlands Antilles: Saba, St. Maarten, and St. Eustatius. These islands are affected by the cold waters of the nearby Atlantic and have a correspondingly diminished fauna. Coral growth here is less exuberant than in the more western areas, and the water is less clear.

However, there is one exciting topographic feature in the deeps off the southwest coast of Saba. Here, kilometers out at sea, lies a vast platform known as the Saba Banks. It never reaches the surface at any point on its broad cap. The Saba Banks, long familiar to local fishermen, were unknown to the outside world until a Dutch Navy hydrographic ship, the *Luymes*, with a group of scientists from the Rijksmuseum, was dispatched to study them. The ship crisscrossed the Banks under extraordinarily accurate radio-computer navigation control and developed a detailed contour map. In addition, samples of marine life from a large variety of sites were collected for study. Almost all of the flat table-top of the structure was found to lie at a depth of 10 to 30 meters. Across the broad cap lie kilometers of sand in a wave-ripple pattern which persists even at depths of 25 meters.

There is noticeable coral development only around the edges of the Banks, particularly on the crest that faces the prevailing current. Here at the edge of the abyss, low-growing corals, encrusting sponges, gorgonians, tunicates, and a wide range of fish species live in the perpetual gloom that results from suspended particles blocking the sunlight. Strong currents move across the ancient sands of the Saba Banks, and large schools of pelagic fish such as swift-roaming jacks and barracuda ply the green waters. The lack of penetrating sunlight and the depth of the water result in limited coral growth. This provides food and shelter for only a relatively small fish population. I was rather shocked by the limited life of the reefs in this section of the Caribbean. Even in the shallow waters inshore on the surrounding islands, the murky water has created an unimpressive reef without the floral character of the warmer, clearer waters of other regions. In particular, the coral growth is concentrated in the large, slow-metabolizing dome corals, with very limited occurrence of staghorn and elkhorn coral. In general, while these reefs contain many of the same species we meet throughout this region, they stand in stark contrast to other, richer zones.

The Bahamas

Finally, we take a brief look at an area which is only marginally Caribbean, geographically, but which retains much of the characteristic fauna of the region. The Bahamas chain consists of some 700 islands stretching from offshore Florida to offshore Haiti. Its weather varies somewhat more widely than that of the rest of the Caribbean, due to its northerly latitude. In the winter the weather frequently becomes quite cold. Fireplaces are lit, and people stroll the long beaches wearing sweaters. These periods of cold, stormy weather are called "vodka fronts," referring to the after-walk activities around the fireplaces. Thus, it is somewhat surprising to find a rich Caribbean reef flourishing in quite temperate water. Some of this effect is undoubtedly due to the nearby Gulf Stream, which carries nearly three billion liters of warmed water per hour through the straits of Florida.

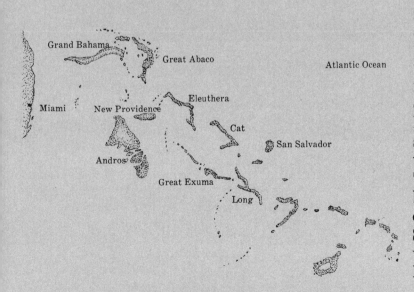

Grand Bahama
Great Abaco
Atlantic Ocean
Eleuthera
Miami
New Providence
Cat
San Salvador
Andros
Great Exuma
Long

One important element of undersea topography in the Bahamas is the Tongue of the Ocean, a tongue-shaped depression between New Providence and Andros islands. Here the water drops abruptly from a depth of 30 meters to nearly 4,000 meters. This precipitous wall has been the site of a number of attempts to set world deep-diving records, some of which were fatal.

The wall has drama, but it is the inshore, shallower reef zones that offer the richest and most plentiful marine life. Wind-driven waves oxygenate the waters along an offshore reef-line about 100 meters from the beach. Along this line, rich stands of staghorn and elkhorn corals provide shelter for a diverse population of grunts, snappers, goatfish, trumpetfish, chromis, damselfish, butterflyfish, and others. Gorgonians are in evidence though not profuse. The flourishing populations of groupers and barracuda bespeak a careful control of spearfishing.

Elsewhere in the Bahamas, at Eleuthera, New Providence, and Long islands, corals and reef life are similarly abundant. One day I was diving off New Providence island at a remote reef quite far from shore, at the eastern edge of the Tongue of the Ocean (Andros being on the western edge). The reef was aswarm with fish, with swift amberjacks making soaring passes up over the drop-off toward us. After an exciting dive we began ascending our anchor line, only to have a hundred or more Bermuda chubs swim about us in a lazy funnel cloud until we were enveloped in a living mass of fish. The sun streamed down through this slowly revolving cathedral until, with great regret, we broke the surface and the spell was lifted.

There are some interesting overall impressions of the Caribbean that the observer takes away. First is the rich diversity of the individual islands despite their common geology and basic fauna. Although the same 48 species of coral have built all of these reefs, the reefs are individually distinguishable one from another. Any reasonably experienced diver could quickly tell a reef of Bonaire from one of Cayman or Roatán. The diver would use certain "signatures" of topography or fauna. Particularly useful would be the sponges of the Caymans, shallow *Acropora* coral beds in Bonaire, and the pillar coral at Roatán.

This is a sea whose isolation is long established, whose biological organization is stable. Any decision by man to disturb that isolation (such as the now abandoned proposal for a new, sea-level Panama Canal) must be carefully considered. Were such a passage opened, the ensuing damage might be irreversible.

The Caribbean exists today as a magnificent, bounteous sea for us to enjoy. Our mistakes of the past are fairly well understood; the underwater wilderness has survived them with only a few haunting casualties. With care on man's part, this colorful and rich sea should be a joy to many future generations.

160. *The colorful spiral feeding arms of the serpulid tube worm* Spirobranchus giganteus. (*Curaçao, Caribbean*)

161. *A solitary barracuda,* Sphyraena barracuda, *patrolling its reef range. It is always curious but only dangerous when theatened.* (*Cayman Islands, Caribbean*)

162. *A graysby,* Petrometopon cruentatum, *full from a recent meal, or perhaps gravid, at rest on the coral at Palancar Reef.* (*Cozumel, Mexico*)

163. *The queen angelfish* Holacanthus ciliaris. (*Cozumel, Mexico*)

164. *A large colony of pillar coral* (Dendrogyra cylindrus), *characteristic of the reefs of Roatán, Republic of Honduras. Some of these colonies reach heights of four meters or more.*

165. *The blackcap basslet* (Gramma melacara), *found in profusion on the deeper reef slopes of Roatán, Republic of Honduras.*

166. *A cluster of sabellid tube worms.* (*Cartagena, Colombia*)

161

162

163

164

165

167. Chromis cyanea, *the blue chromis, found hover-ing in great numbers above many Caribbean reefs, particularly off Curaçao and Bonaire.*

168. *A Caribbean sea whip, the most common gor-gonian in Bonaire. The reef shallows, some 25 to 50 meters off the beach, contain forests of these organ-isms, slowly moving in the wave-surge.*

169. *An exquisitely frilled Caribbean nudibranch.* (*Curaçao*)

170. *An azure vase sponge. The tentacles are those of an ophiuroid (brittle starfish) whose body is hidden for protection. Only at night do the ophiuroids crawl openly on the sponges to feed.* (*Cayman Islands, Caribbean*)

171. *The common jellyfish* Aurelia aurita *floating in open water.* (*Curaçao, Caribbean*)

172. *The frogfish* Antennarius multiocellatus *taking flight from a coral head. This remarkable anglerfish uses the lure mounted on its nose to entice small fish toward its large mouth.* (*Curaçao, Caribbean*)

173. *The smooth trunkfish* Lactophrys triqueter. (*Cozumel, Mexico*)

174. *A triton trumpet,* Charonia tritonis, *plunging its proboscis into a captive sea urchin. It takes the triton trumpet about an hour to remove all the flesh. Skin, bones, and spine are left.* (*Curaçao, Caribbean*)

175. *The jackknife fish* Equetus lanceolatus. (*Curaçao, Caribbean*)

176. *Tentacles of a pink anemone found off the Netherlands Antilles in the Caribbean.*

168

169

170

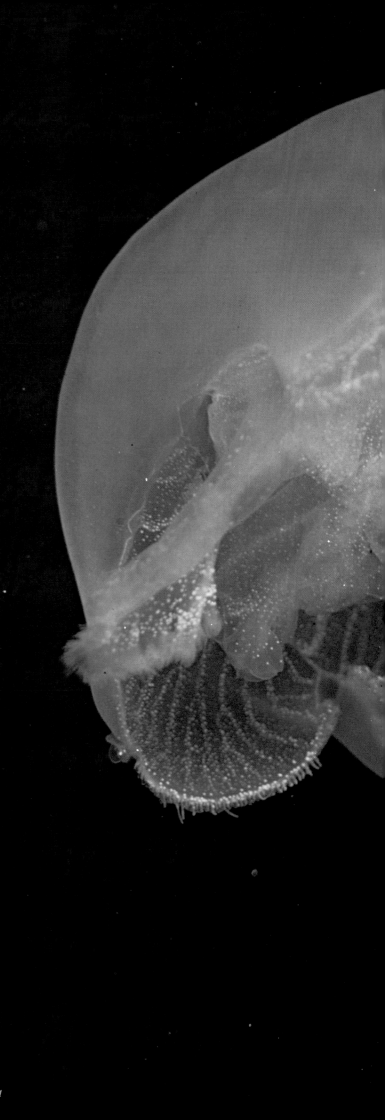

171

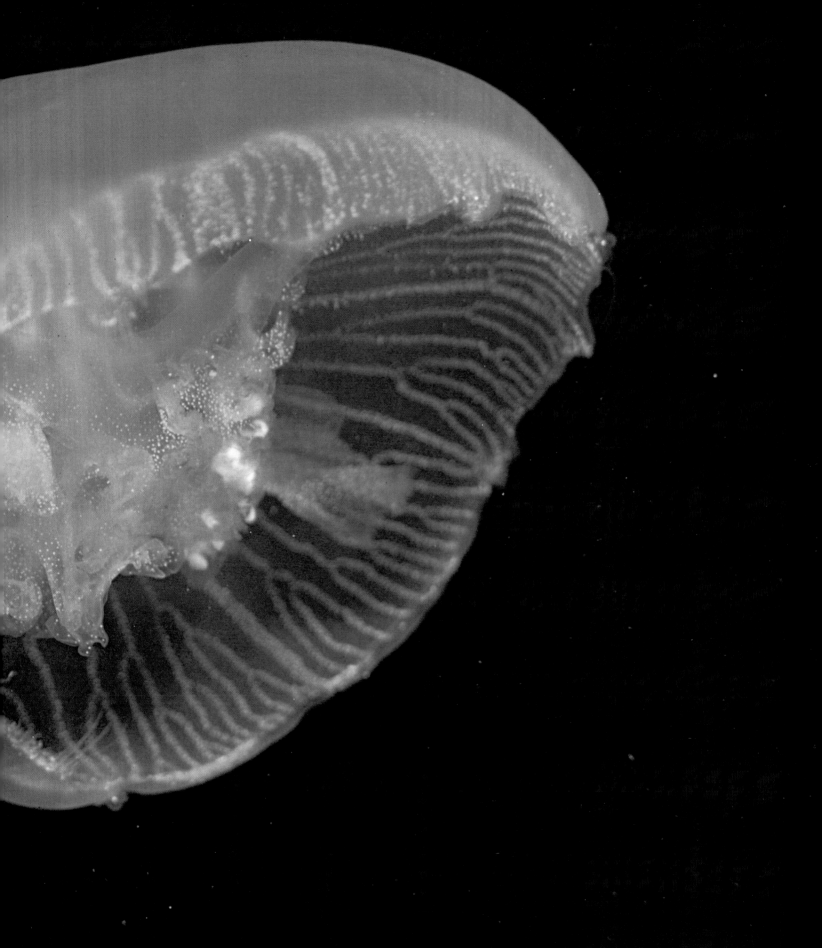

174

175

176

177. *A gorgonian sea fan. (Caribbean)*

178. *An azure vase sponge that has grown upon the skeleton of a black coral tree, situated on a precipice wall. Beneath the wall, the water is over 100 meters deep. (Cayman Islands, Caribbean)*

179. *Another of the incredibly varied sponges of the Cayman Islands. This one, too, has grown upon the framework of a still-living black coral (antipatharian) colony.*

180. *A delicate vase sponge, photographed on a reef some 15 meters beneath the surface off the coast of Cozumel, Mexico.*

181. *A young diver silhouetted behind a large gorgonian on the precipitous North Wall of Grand Cayman Island. (Caribbean)*

182. *A dainty Phoronid feather worm that grew in a coral crevice. (Saba, Caribbean)*

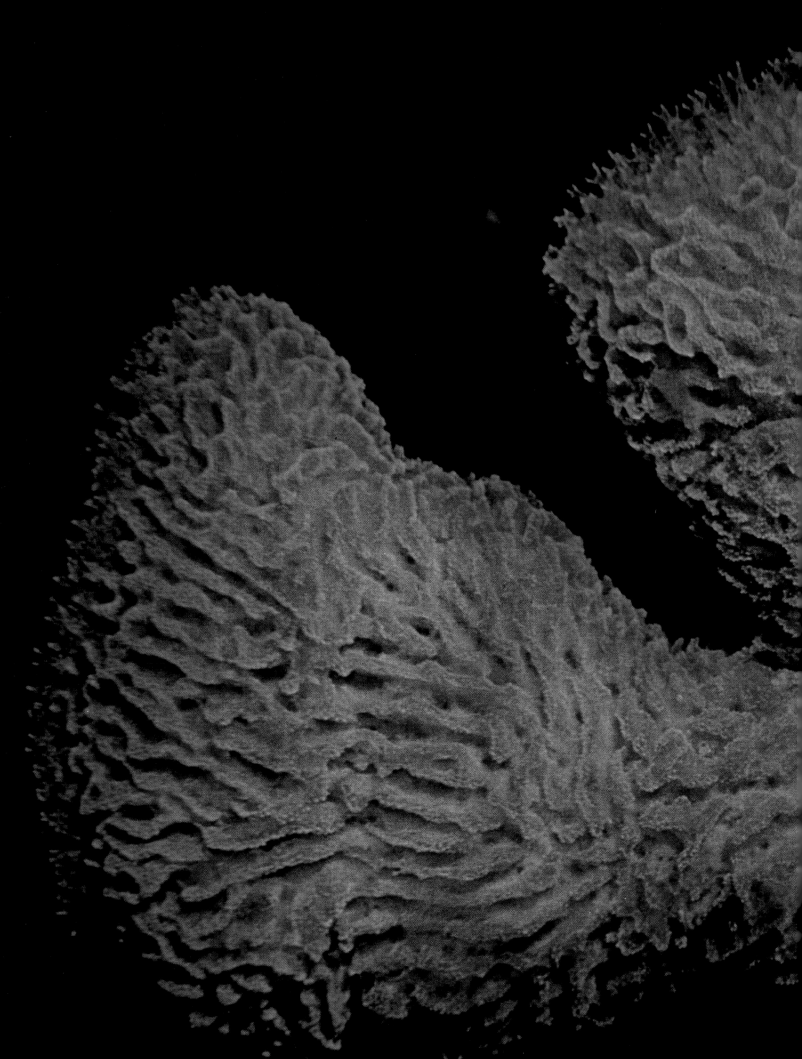

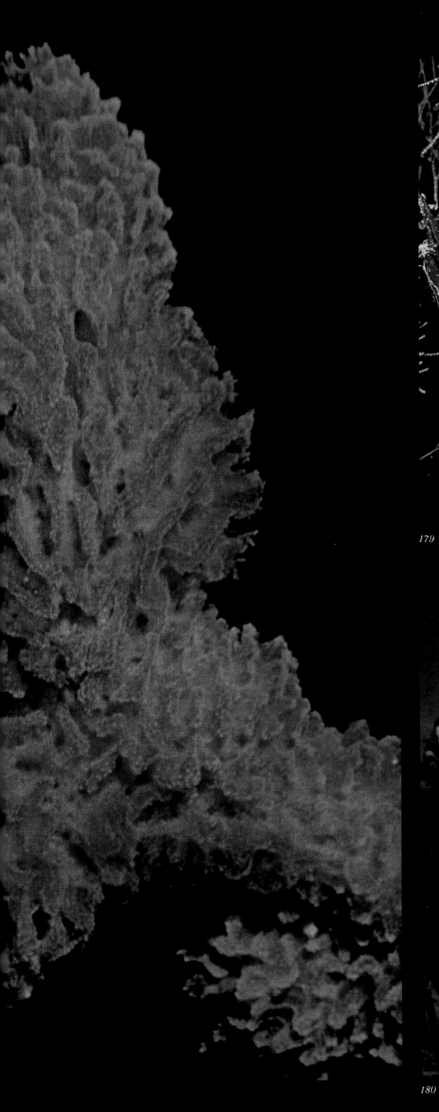

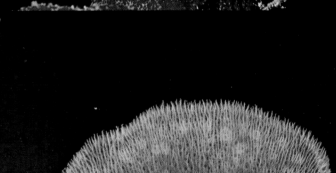

179

180

The Fringe of the Tropics : Hawaii, Baja California, and the Galápagos

The three regions in this chapter are geographically distant from one another, although all are in the eastern portion of the Pacific and on or north of the equator. All three were born in the crucible of volcanic eruption and lie about the fringes of the vast tropics of the South Pacific. Unlike their neighbors to the south and west, these regions have been affected in the development of their marine fauna by cooler water from polar regions. These cold currents were described in the first chapter as moving along the western coasts of the continents toward the equator. We would expect these regions to develop differently from the truly tropical islands, and indeed they have.

Hawaii Rising starkly from the cobalt-blue abyss of the Pacific, the islands of Hawaii leave no doubt as to their origin. They are not only volcanic, they are also geologically the youngest of the three eastern Pacific regions. Two of the volcanoes in these islands are over 4,000 meters in height and show relatively little erosion of their conical forms.

The Hawaiian Islands stretch across 3,000 kilometers of ocean. They were apparently settled in about 750 B.C. by mariners from the area now called French Polynesia. Various legends trace the original ancestors to Huahine, Havaiki, or even Bora Bora, thousands of kilometers to the south. Regardless of their origin, the Polynesians in Hawaii maintained their culture in isolation for just over a thousand years. They possessed a sophisticated understanding not only of tides, winds, and currents, but also of the lives and habits of the multitude of creatures in the sea about them. Fish, for example, played a powerful role in their highly personal religion. Each fisherman had a personal deity among the fish, and to this day there are still small shrines where sacrifices to godfish are made.

The first European known to have visited the islands was a Spanish captain, Juan de Gaetano, in 1555. Spain never claimed the islands, however, and so the Hawaiians continued in their isolation until the arrival of Captain James Cook in 1778. Cook was welcomed by the Hawaiians, who offered cheerful assistance to the British explorers. A year later, however, Cook was killed at Kealakekua Bay in a skirmish over a stolen smallboat. This tragic loss of Britain's greatest mariner also cost the British the possession of these strategic islands. By the time their next ship called in 1789, King Kamehameha had begun a stable and powerful dynasty which would last a century.

Polynesian culture was finally overthrown not by arms but by the misguided efforts of missionaries. Crushed between the disruptive methods of the zealous missionaries and their contagious diseases, the Polynesian population, culture, and society fell. Some of the missionaries ultimately became planters, and their families, with tremendous landholdings, shaped much of the islands' modern history. Succeed-

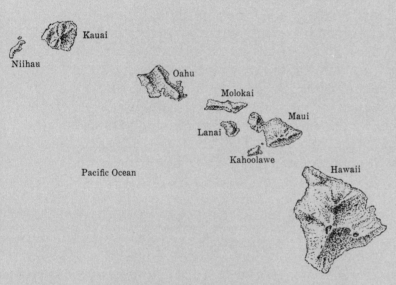

The Hawaiian Islands. This is a chain of volcanic islands that stretches across 3,000 kilometers of the Pacific Ocean.

ing waves of Chinese, Filipino, Korean, German, and Portuguese settlers, and more recently hordes of tourists and investors, have only further enriched these early families.

The underwater wonders of Hawaii are radically different from those of the Caribbean. The waters are several degrees colder, cooled by the fringe of the California current flowing down from the Arctic. Hawaii's location, relative to its cool waters and larvae-bearing currents, has resulted in rather impoverished coral reef development, though its tropical fish population is profuse.

The island chain's volcanic activity, the sudden quenching of flowing hot lava by sea water, has shaped the underwater topography. Along the drowned volcanic flanks of these scattered islands are soaring undersea arches, caverns, and tunnels of lava. None of these are similar to the reef-coral construction of the Caribbean.

Around these unusual lava structures of the marine terrain, with their sparse coral covering, swims a rich array of fish life. For the diver who has done his early diving in the Caribbean, the corals of the Hawaiian Islands are a disappointment. However, their colorful array of butterflyfish, surgeonfish, and tangs are brilliant compensation.

Off the big island of Hawaii there is a particularly fine underwater site at Kealakekua Bay. In the shadow of a monument to Captain Cook, divers enter a semi-enclosed bay teeming with fish and invertebrates and marked by the most prolific growth of small coral heads in all the islands. Swarms of rudderfish (*Kyphosus cinerascens*), angelfish (*Holacanthus arcuatus*), damselfish (Pomacentridae), parrotfish (Scaridae), wrasses (Labridae), and surgeonfish (Acanthuridae) swirl and flow about the diver. On the low coral heads sit brilliant red slate-pencil urchins (*Heterocentrotus mammillatus*), their stout spines sure protection against attack (Illustration 183). Other urchin species, not as well armored, huddle beneath the sheltering edges of the coral awaiting nightfall before they venture out.

The entire leeward coastline of the big island, the Kona Coast, is a rare underwater wilderness. Generally the coast is composed of rocky cliffs of black basalt which create spuming blowholes and spray when waves crash against them. At the base of these cliffs are wave-eroded caverns and arches. The underwater terrain then slopes outward in parallel ridges and gullies to an eventual drop-off. Much of the good diving along this coast is in quite shallow water, at 3- to 15-meter depths.

Off the coast of the island of Lanai lies a series of large subsea caverns, poetically and aptly named the Cathedrals. The name is especially appropriate in one magnificent cavern which is perhaps 20 meters long, 10 meters wide, and almost 10 meters high. Entering the arched "rear" opening to the Cathedral, one gazes in awe at the luminescent spectacular at the other end. Several openings allow the sun's rays to pour into that small portion of the cavern, accentuating the empty darkness of the rest. The effect is stunning, a rare and unforgettable scene.

It was outside the Cathedrals that I had a particularly memorable encounter with a marine creature. The huge, rocky cavern structures are separated by sand-filled canyons. Here, emerging from the vaulting gloom of one of the Cathedrals, I discovered an eagle ray peacefully browsing above the sand. An eater of mollusks, the ray exposed its prey by raising a cloud of sand through fanning with its two-meter wings. I swam closer, blazing away with camera and strobe light. Suddenly aware of my approach, the ray lifted its head from the sand and peered at me for a moment like some grizzled hermit startled in his cave. Then as if stung, it soared swiftly away, almost disappearing at the far end of the long valley.

But then it turned. As my heart pounded and I recited a litany of exposure values, this graceful creature winged its way along the edge of the coral cliff above me, passing perfectly beneath the sun. As I saw its wing nearly touch the sunburst overhead, I felt a moment of rare communion. It seemed no longer afraid, and had flown over almost like a pilot dipping his wings in farewell.

Several kilometers from Lanai, in the straits between Maui and Kahoolawe, the very top of a small volcanic cone juts from the deep water. More than half of the crater rim is gone now, victim of the ever-punishing waves or perhaps of an interior collapse. The surviving segment of the rim slices above the sea surface like some rocky half-moon. This is the crater of Molokini, whose waters abound with the false moorish idol (*Heniochus*), lazy manta rays, barracuda, jacks, sharks, and rudderfish.

Along the shallow rim of Molokini, occasional strong currents urge divers to caution, but all is quiet on the volcano's deep flanks. At one spot on the collapsed side I have encountered a huge school of *Heniochus* near a series of ledges under which three white-tipped sharks were resting. At such moments, when two or more exciting subjects are performing simultaneously, a photographer's concentration approaches overload. Each subject invariably involves a different strategy, perhaps even a different camera, but the cameraman is greedy to document everything. Sometimes, as in this case, the choices are extremely difficult.

A final important attraction in these islands is a marine spectacular placed by man. In the ship channel off the oldtime whaling town of Lahaina, Maui, lies the American submarine *Bluegill* sitting upright with its keel resting on a flat sandy bottom 45 meters deep. The sail soars upward to within 20 meters of the surface. The *Bluegill* is so recent an underwater addition that only a limited marine growth has appeared on her exterior. Even so, close examination reveals the tubes of countless marine worms on the once sleek metal. These worms are usually the first settlers on any new metal or glass surface in the sea, followed later by algae, corals, and sponges. Around the submarine's vast bulk hover schools of goatfish (*Mulloidichthys samoensis*) and occasional jacks. Many years from now the slow accumulation of marine organisms on the submarine should create a reef, to which increasing numbers of small and large fish will come. The *Bluegill* was sunk to create a training ground for Navy divers, but the long-range result will be an entirely new undersea community.

In the coming years, Hawaii's reefs may get more pressure than they can bear. The pace of development and tourism could severely threaten coral areas such as Kealakekua Bay. But action is now being taken to protect some of these underwater wildernesses. For example, a national park now being planned here should reduce spearfishing and linefishing. Protective measures must be established now before such activities as the collection of shells and corals for tourist souvenirs completely unbalance the marine ecosystem.

Baja California and the Sea of Cortez

Probably the least known underwater wilderness in the world lies a mere 80 kilometers from the United States–Mexico border. Here in the Gulf of California lies the youngest and in some ways the richest of the world's seas—the Sea of Cortez.

Some 10 to 15 million years ago, in a cataclysmic geologic event, an enormous coastal area of Mexico was suddenly ripped away from the mainland. This upheaval along the infamous San Andreas fault tilted what is now the Baja California peninsula westward until an enormous chasm yawned between the two land masses. We can only imagine the cataract force with which the mighty Pacific roared into the resulting crevasse. Entire volcanic peaks ripped loose and slid into the yawning trench, forming what are now chains of coastal islands scattered through the Gulf of California.

The depth of the sea thus created exceeds three kilometers in some places, and its width ranges from 50 to 250 kilometers. Its waters are quiet with a sunbaked upper layer, due to the protecting land masses on three sides.

The Baja peninsula, a 1,500-kilometer area of mountain, desert sands, and cactus, is a haven for a wide range of hardy terrestrial species of plants and animals. In fact, its isolation has been so complete that bird species long thought extinct have been found in its immense wastes. But it is in the vast deep waters which the peninsula shelters that a miracle of life occurs on a gigantic scale.

This wellspring sea has been known since the Spanish Conquest as the Sea of Cortez. Named after Hernando Cortez, the conqueror of Mexico, the Gulf was first probed by one of his scouting expeditions. The scouts returned with tales of pearls and riches. Beginning in 1535, Cortez and his captains explored the

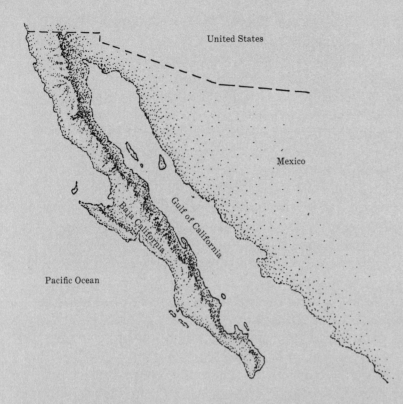

Baja California and the Sea of Cortez. Also called the Gulf of California, the Sea of Cortez contains perhaps the richest concentration of free-swimming life found in any single ocean.

sea. One of his men, Francisco de Ulloa, reached the mouth of the Colorado River at the northern tip of the Gulf in 1539. These early explorers were followed inevitably by zealous missionaries and pirates who, each in their own way, destroyed the Indian cultures they encountered. Only the enduring land and sea survived those few centuries.

The Sea of Cortez contains perhaps the richest concentration of free-swimming life found in any single ocean. Incredible fish shoals sometimes kilometers long churn the surface as they feed on smaller fish and the immense plankton population of the Gulf. This prodigious concentration of plankton results principally from the action of Pacific tidal waters. The Sea of Cortez has very little tide of its own because it is so narrow, but the nearby Pacific rises and falls as much as two meters. Thus the tidal waters of the Pacific that enter the mouth of the Gulf near Cabo San Lucas affect a 1,000-kilometer sea in a mere five and one-half hours. The obvious result is a massive and swift movement of oceanic water that pours vast amounts of nutrients into the lower Gulf twice each day. The presence of the concentrated plankton in the sun-baked and calm surface waters produces fish food, and thus fish populations, in staggering profusion: whales, porpoises, rays, tuna, jacks, and dozens of other open-water predators. Here is a textbook example of the oceanic food pyramid. The abnormally high concentration of phytoplankton (plant life) supports an enriched population of zooplankton (animals), which are the prey of pelagic crustaceans, fish, and squid. These in their turn are prey to schools of larger herrings, anchovies, mullets, needlefish, and others ranging from a centimeter or so to several centimeters in length. These are the prey of swift, voracious tuna, mackerels, sierras, giant needlefish, and sharks.

When natural conditions favor intense plankton growth, a burgeoning always follows among predator species. The imperatives of biology assure overall balance. In the Sea of Cortez, the result is a legendary wealth of pelagic fish and mammals. Boatmen describe leaping porpoises in schools of thousands. Fishermen report the ocean surface frothing from the frenzied feeding of countless swift-running hunters. Divers recount the awesome experience of hovering at a depth of 30 meters as hundreds of close-packed hammerhead sharks pass in silent formation overhead. Perhaps the pinnacle of human experience with the ocean life here is moving on the open water with an escort of finback or sperm whales. These giants, which reach a weight of 80,000 kilograms, move placidly through the Gulf, occasionally keeping pace with small boats.

Despite the incredible abundance of pelagic life here, the Gulf remains subject to immutable laws concerning coral temperature tolerance. The frequent cold temperatures in Gulf waters (when the cold deep water is stirred by storms, seasonal variations, or abnormal tides) practically prohibit reef-building corals here. There are a few clusters of night-blooming coral in bright colors but none of the warm-water, reef-building species. A number of different branching gorgonians, related to corals, flourish in the colder waters below 10 meters. But most of the shallow-water scenery consists of fallen boulders, some of them immense and covered with algae and scattered sand.

Thus to the snorkeler and diver the waters of the Cortez offer a paradox: rich fish life amid rather modest reefs, huge schools of goatfish, butterflyfish, grunts, and snappers patrolling valleys with only scattered corals visible. In the crevices between the huge boulders, wary groupers and spiny pufferfish watch for predators and prey. Now and then a school of gray surgeonfish flows past, or a turtle, or a shark. On the boulders here and there a colorful starfish or sea urchin accents the barrenness. In the shadow of a boulder, a giant hawkfish is hidden for a moment by its patchy camouflage.

In the waters of the southern tip of Baja, there is frequently a temperature difference of 10 to 15 degrees between the surface and a depth of 30 meters. The water temperature changes abruptly in layers. The borders between these layers, called thermoclines, are an unpleasant shock to the human system but are intriguing nonetheless. A thermocline of 5 degrees Celsius often has fine sand particles floating visibly upon the denser cold water.

In the icy dark waters below 30 meters, brilliant gorgonians blooming in blood-red and white seem out of place on the harsh rock substrate. But this very sharpness of contrast lies at the heart of this sea's fascination to visiting scientists and divers. On sand shelves in this cold zone one may encounter the shovel-nosed shark, while large hammerhead sharks drift by in the darkness. Small and large manta pass, and large schools of cornetfish and spiny pufferfish suddenly appear, hovering above the sun-gleaming sand. In Baja California, a new and yet ancient land has now been opened to the automobile, the camper van, and the motorcycle. The harsh terrain, long silent, now echoes to the throb of engines, the whine of swift vehicles. We have found the treasure. Will we destroy the prize even in the finding of it?

The Galápagos Islands

The islands of the Galápagos are stark volcanic rock which burst from the sea during a period of seismic activity some three million years ago. Officially a territory of the Republic of Ecuador, the islands lie scattered across 25,000 square kilometers of ocean on the equator, 1,000 kilometers due west of the port of Guayaquil. There are 13 major islands, six minor islands, 42 rocky islets that carry official names, and a host of further rocks and islets that bear only local names.

The islands were discovered in 1535 by the Bishop of Panama, Tomás de Berlanga, when a ship on which

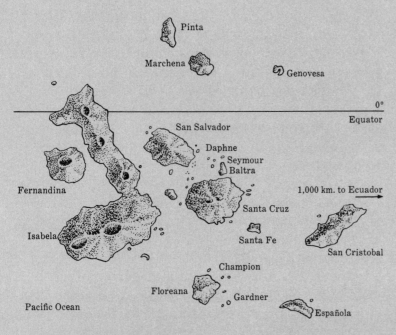

The Galápagos Islands. The cold currents around these rugged equatorial islands support a spectacular marine fauna.

he was voyaging to Peru was carried off course by unexpected currents. He wrote of the islands in a report to his emperor, Carlos V of Spain, making special mention of the giant tortoises (*galápagos*), the iguanas, and the birds, which were surprisingly tame.

In the years that followed, Spanish navigators charted and named the islands. These explorers were followed in the 17th century by English pirates who gave English names to the same islands. Even today, each island has two names. Santa Cruz Island, site of the Charles Darwin Research Station, is also officially Indefatigable Island; Isabela is also Albermarle; Genovesa is Tower, and so forth.

After the pirates faded into history, the islands were visited by whalers, who slaughtered the stately tortoises for food and sport. Worse, the goats and pigs they brought to the islands, as well as the rats that came off their ships, went feral. This introduction of efficient competitors among noncompetitive local species put severe pressure on such animals as the tortoises. The goats, highly efficient foragers, completely cleared out the foliage on which the tortoises fed. Since a goat can reach higher than a tortoise, many of the large reptiles starved to death looking up at rich greenery just beyond their reach.

The most profoundly influential visitor to these volcanic islands—long known as Las Islas Encantadas, or the Enchanted Islands—was undoubtedly the 26-year-old naturalist Charles Darwin, aboard *H.M.S. Beagle*, in 1835. Observations that Darwin made of Galápagos species were vital in directing him to the fundamental conclusions published in his epochal book *Origin of Species* (1859). Perhaps the most important observation was that there were 13 species of finches on the islands, with up to 10 species on a single island. Darwin found that the principal difference in the various species lay in the shape of their beaks, with different beaks denoting different feeding patterns. Departing dramatically from the prevailing theories of the independent creation of all species, Darwin postulated the theory of a single common ancestor developing rapidly into local variants in the competitive vacuum of these isolated islands. These later variants, he concluded, were the result of competitive pressures for food. Different species had been forced to adapt to local conditions as they spread across the islands, developing varying food-gathering techniques.

In the cold green waters that surge about the remote and rugged Galápagos thrives a fauna as spectacular and varied as the celebrated terrestrial animals of the islands. Because of a lack of funds, equipment, and knowledge, very little underwater exploration has been undertaken in the archipelago. Even local boatmen have been limited in their ability to probe the vast and diverse undersea terrain. Only completely self-contained expeditions in good-sized vessels could offer more than the barest glimpse beneath the surface. For these reasons the Galápagos have never been considered a great site for diving, nor have their underwater fauna enjoyed the attention accorded the terrestrial creatures.

More recently there has been scientific research into marine species in the islands, and the vast oceanic movements which brought the Galápagos their land fauna are now known to be the source of their marine life as well. For example, scientists now know that the Humboldt Current acts as a life-bearing stream flowing north from Antarctica to the Galápagos. This cold water influences the archipelago from June through December. The current is undoubtedly responsible for the penguin population of the islands; its influence alone renders these equatorial outposts hospitable to these birds, found chiefly in Antarctica. The second major current around the Galápagos comes from the northeast over cycles of several years. Locally known as El Niño, this stream wells up in the Gulf of Panama and occasionally runs southward as far as the Galápagos, bringing warm water and life forms to the archipelago.

A third major stream periodically influencing the Galápagos is the Cromwell Current. Originating in the far Pacific and as big as the Gulf Stream, this flow moves slowly eastward, upwelling 100 kilometers east of the islands to bring great food riches from the deeps.

On these great currents both the refugees which long ago became Galápagos parent species and more modern immigrants were brought to their island home. Identified inhabitants comprise 289 species from 89 families, of which 77 percent are shallow-water forms. Only 7 percent of the species seem to have originated in Peru or Chile, while 54 percent are eastern Pacific species such as those also found at Baja California and along the western coast of Central America. This would indicate perhaps a higher survival rate of species transported from the relatively nearby Gulf of Panama as opposed to those from the remote Antarctic. Twelve percent of the species are from the Indo-Pacific, and fully 23 percent are peculiar to the Galápagos, having evolved in isolation from some ancient species. A few species are of unknown origin, and others undoubtedly have yet to be discovered and classified.

Until now, scientists alone knew about the undersea natural history, currents, and statistics. Tourists in the islands have for years visited colonies of birds, tortoises, lizards, and other creatures but have never seen what was beneath the surface of the sea. Amateur diving was simply unknown. During recent years an increasing number of scientific and sport divers have begun to visit these enchanted islands. The waters of the Galápagos are rich in marine life, but one of the sobering conclusions of my experience there is that they are far from inexhaustible. Even as many land species are dwindling or facing extinction, the creatures of the sea are under attack from commercial lobstering and fishing interests. As in other tropical areas that report overfishing and declining catches, government officials in the Galápagos are only slowly beginning to act to protect their marine fauna. Even though recent laws prohibit any vessel of foreign registry from visiting the islands without permits (which are very rarely given), the waters of the Galápagos still await the National Park designation long ago extended to the islands themselves. Prohibitions against foreign boats do not affect Ecuadorian fishermen. It has been reliably reported that over 200,000 lobsters were taken from one group of islands in a single season by Ecuadorians.

The Galápagos marine ecosystem, like any other, cannot long tolerate this level of depredation without irreversible damage. The prospect is especially tragic in these islands. In variety and unique forms, the marine fauna rival the justly famous land animals. The extraordinary 23 percent of the shore fishes that occur around these islands and nowhere else in the world must be considered as being as valuable a treasure as the tortoises, boobies, and albatrosses of the terrestrial National Park.

The underwater vistas themselves are equally diverse. Each diving location has its own characteristics and creatures, and they are so different from one another as to seem to be in separate oceans, or even on different planets. For example, Daphne is a volcanic cone jutting from the sea near Baltra, site of the only Galápagos airport. Rising from a cobalt-blue sea, the volcano's scarred brown symmetrical flanks pose the starkest possible contrast with the surrounding waters, representing a momentary standoff in the endless war between the all-encompassing Pacific and these lonely lava outposts. Tourists visit Daphne because of its incredible crater full of blue-footed boobies. Yet at the very spot visitors are set ashore, they pass over a remarkable marine wilderness. In 10 to 20 meters of water live swarms of angelfish, wrasses, starfish, and club-spined sea urchins, as well as moray eels, groupers, snappers, sharks, scorpionfish, and other species, many of which are to be seen nowhere else. The creatures are unique in both color and size. I observed one massive scorpionfish double the length of any I had ever seen. The waters of the Galápagos are fairly warm—generally in the mid-20s on the Celsius scale—during the short spring months, but are down to the high teens by July. Visibility does not particularly vary from season to season, but there are differences in water clarity between the northern islands like Tower and those in the southern quadrant such as Floreana and Champion. Another good diving area is located off the rocky flank of Isla Seymour and Isla Mosquera. Once as I drifted along the coast of Seymour in a fairly strong current, I looked up toward the surface and caught sight of an enormous porcupinefish

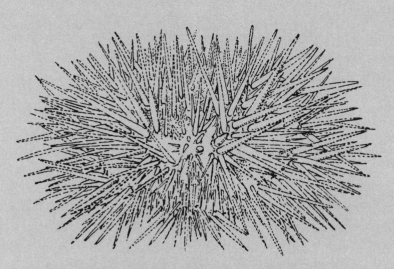

Rock-boring sea urchin. These animals find protection in reef crevices from wave splash.

(*Diodon hystrix*). Readying my camera I gave frantic chase, only to be reduced to helplessness and frustration as the current carried me irretrievably away.

At the end of this dive along Seymour lies Isla Mosquera, a low sandbar inhabited by a good-sized colony of Galápagos sea lions (*Zalophus californicus galapagensis*). Those who have never dived with sea lions can hardly be prepared for the circus they provide. Throughout every dive, squadrons of these playful mammals swirl about, rolling their eyes and cocking their heads interrogatively, corkscrewing their bodies gracefully in contrast to their clumsiness on shore. There seems no question that the limpid-eyed females try to communicate with divers. They will swoop by and pop a burst of bubbles, almost as if imitating the exhalations of the divers. The more one dives with sea lions, the more one tends to view them as companions.

For a totally different underwater experience there is Gordon Rocks, with its sheer pinnacles thrusting up from the depths. These peaks, standing far from shallow waters, offer no anchorage to a dive boat. The boats stand away from the rocks, and the divers must swim to them. Miscalculation here can be dangerous, as currents can suddenly sweep a diver past the rocks and make it necessary for him to be retrieved by motorboat.

The pinnacles of Gordon Rocks are a structural extravaganza. The rock plunges on all sides to depths of 100 meters or more. Its face is covered with holes, each filled with a sea urchin. The urchins bore the holes with file-like rasps to afford themselves protection against predators during the daylight hours. Around the sea urchin holes are clusters of night-blooming coral in various shades of yellow and rose, tiny floral clusters over the abyss (Illustration 139). As you dive along this precipitous wall, the water around you is filled with sea lions, or sometimes with majestic amberjacks that reach a length here of nearly two meters (Illustration 211). The amberjacks soar swiftly in the green water, sweeping from darkness to darkness, sharing with the divers a brief moment of illumination before they race off into the depths. On one dive my partner and I left the ghostly amberjacks behind and rode the current sweeping past the rocks, picking up as we went a brilliant swarm of white-striped angelfish (*Holacanthus passer*). In a few moments we had been hurtled through the pass, into calm water on the lee side of the spires. We photographed the angelfish and the ever-present sea lions briefly, then were amazed to see that the great shoal of amberjacks had followed us through. In the Galápagos, either in the sea or on land, the wildlife seems to be as interested in human visitors as the tourists are in the wildlife. Fear of man has not yet reached the islands, and we can only hope that it never does. In the northern Galápagos lies the remote, stern Isla Genovesa (Tower), home of a large colony of frigate birds. In the late spring months the frigate birds can be found courting and mating, the males with their huge red gular pouches inflated like balloons on their chests. The awkward chicks constantly squall to be fed, and adults back from the hunt regurgitate a share of food beak-to-beak with the hungry chicks.

Genovesa's bay is the caldera of a sunken volcano, with only one narrow passage through the forbidding black lava walls to the open sea. The water is icy cold and extremely murky. Yet under the surface the great blocks of lava are encrusted lavishly with orange coral polyps, algae, and sponges in a wide range of pastel colors. Great herds of gray surgeonfish materialize out of the gloom, holding formation as they wheel and bank.

Another superb diving site is Isla Cousins. Here as at Gordon Rocks no anchorage is possible, so one swims from a passing boat to the rocks, accompanied by the inevitable squadron of sea lions. Cousins was amazingly rich—at one point I had not one but three large Panama horse conchs posed on a rock ledge. Since these lushly-colored creatures would withdraw into their shells when moved, I would shift one to where I could photograph it and then leave it in order to photograph the rest. Invariably when I returned, the first conch would have relaxed and emerged into full view.

Here, too, I made a mistake that almost proved costly. Laying one of my cameras on a ledge, I photographed with another. Moments later I looked up to see a sea lion attempting to pick up the large rig in his teeth and carry it away. Had he succeeded he would have dropped it a short distance away into several hundred meters of dark, cold water. These curious friends cannot be trusted with such temptations.

Also at Cousins in deep water is a vast bandshell of a cavern lined with prolific growths of colorful gorgonians. There I found a pair of large, yellow-eyed moray eels with savage-looking fleshy horns. These aggressive morays snapped and glowered at the brilliant flash of my strobe light, which must have been an unaccustomed intrusion into the never-ending darkness of their world.

A totally different segment of the Galápagos marine fauna is found at Isla Bartholomé. Here a sloping sandy bottom is home for curious pufferfish and pink sea urchins which camouflage themselves with bits of debris impaled on their spines. Here also is found the unusual red-bellied batfish (*Ogcocephalus darwini*), poised watchfully to pounce on any tiny prey that come to investigate the lure he dangles from the tip of his nose. Here one may also encounter a flock of penguins, the shy, fast-swimming birds of the Antarctic. There is a certain fascination in knowing that each day you are steaming back and forth across the equator, and yet can anchor next to rocks covered with docile sun-bathing penguins.

By far the largest of the "enchanted islands" is Isa-

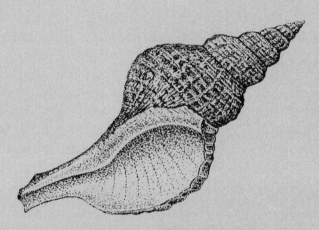

Shell of a Panama Horse Conch. When alive the fleshy tissue is a brilliant blue and flaming red.

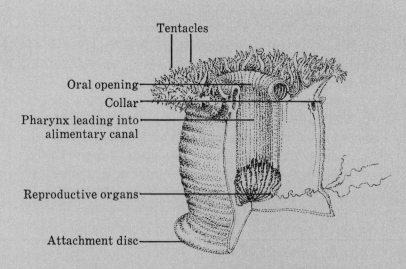

Cross section of a sea anemone. When viewed from above, all sea anemones are radially symmetrical.

Labels on illustration: Tentacles; Oral opening; Collar; Pharynx leading into alimentary canal; Reproductive organs; Attachment disc

bela. Here five major volcanoes jut into the bright sky. The most impressive is Alcedo crater, whose gigantic rim is 40 kilometers around. A few kilometers to the north is Punta Vicente Roca. This abrupt peninsula whose surrounding waters sparkle brightly in the sun is scenically spectacular. To the right as one anchors is a lava cliff rising nearly 1,000 meters into the brilliant sun. To the left, in the channel between Isabela and another island, Fernandina, pass occasional finback and sei whales placidly migrating north. Nearer the boat, huge ocean sunfish (*Mola mola*) bask on the surface. Directly ahead, not a hundred meters away, is a vaulted cave extending far back into the solid lava.

I was not successful in filming the sunfish and whales from our chase boat and was drawn back to the ominous-looking cave. It was unbelievable. The roof is five to ten meters above water level, and the floor perhaps ten meters underwater. The walls are profusely gardened with brilliant gorgonians, black coral, anemones, and colorful coral clusters. The floor is littered with lava boulders around which mill wrasses, snappers, groupers, and other good-sized fish. The wave-surge pulses slowly in and out of the cavern. One burst of exciting photography featured an unforgettable underwater encounter with a swift-swimming, flightless cormorant, skimming among the rocks looking for a meal. Then I moved deeper into the cave.

I did not realize that a dozen young sharks made their home in this cave, and that my intrusion was gradually cornering and exciting them. After a while menacing shadows began to race by on all sides, and for just a moment man's deepest fears reached me with an icy touch. Then the sharks were gone, and I completed my filming of the cave. Later, when I returned to the dive boat, I took one of the skiffs and drove it slowly about on the mirror-smooth surface water inside the cavern. The silent sentinels had returned to their slow sweeps, easily visible in the clear, still water near the rear of the cave.

Tagus Cove is a box-like bay with cliffs towering on three sides. It contains some of the coldest water and the richest life in the islands. Forty meters below the surface in the cold near-darkness bloom families of brilliant white anemones, while on the lava walls are plush carpets of gold and rose-colored colonial corals. Here and there a spectacular *Heliaster*, a many-armed starfish (Illustration 200), sparkles like a jewel against the stark rock, while schools of varied fish pass on their regular rounds.

One night my group took the skiff to a remote spot far from our main anchorage and began diving. In the sea, nocturnal life forms are everywhere different from those encountered by day. I found this to be especially true when night-diving at Isabela. The night-blooming corals were as lushly hued as any I had ever seen, and such unusual nocturnal denizens as slipper lobsters, a horned shark, and a golden seahorse came to pose for our cameras.

On our way back to the anchorage in the outboard-powered skiff, we found ourselves skimming under a sky full of blazing stars. The boat's wake glowed with phosphorescence caused by tiny plankton whose response to being disturbed is a burst of chemical light. It was a breathtaking experience, as if we were riding a comet through the endless reaches of space. By the time we returned to the main boat our curiosity had been aroused, so we put a light over the side and collected buckets of the tiny creatures. Delicate shrimp, crabs, and other forms darted around in the water—a roiling stew of life. It was easy to understand how these waters supported such a host of larger forms; the food chain is very solidly based.

One site in the Galápagos inevitably is the highlight of all expeditions. Punta Espinosa, on the Isla Fernandina, across the channel from Tagus Cove, harbors pelicans, hawks, herons, and a major colony of marine iguanas. These iguanas and the great land tortoises are the hallmark species of the islands, known throughout the world. On one trip we practically lived in the water with the iguanas, watching them feed, swim, and bask in the sun. The manhandling the iguanas take from the omnipresent sea lions fascinated me. I was filming a reptile placidly browsing about three meters under the surface when a sea lion zoomed by and bowled over the iguana apparently for the sheer pleasure of it. Another time one of us observed two sea lions tossing an iguana back and forth on the surface like a rubber ball. Throughout all such harassments the iguanas displayed a stoic indifference, regaining their dignity and resuming their feeding as though the incidents had never occurred (Illustrations 204–207).

On Isla San Salvador (James) lives one of the most appealing sea lion colonies in the Galápagos. The animals swarm in the surging waters near the island, protected from deep-water predators by a network of still, clear-water canals in the lava rock. Here in the canals at sunset these fur sea lions slowly cavort, lolling about on the surface and rolling their large eyes before settling snugly into rocky crevices for the night. In the water by day they have a beguiling pose: hanging head-down from the surface, they swivel this way and that to watch their visitors. Periodically they will dash back and forth like excited puppies, only to stop in front of a visitor and hang upside down once more.

The islands of the southern sector offer some of the most benign conditions for diving in the Galápagos. The waters are somewhat warmer and clearer than in the north. Coral development is exceptional for the Galápagos, and there are massive concentrations of sea lions; fish (among them damselfish and giant hawkfish); and other marine species, including large green sea turtles, brilliantly colored nudibranchs (shell-less snails), large white stringrays, and sponge-coated sea urchins.

Among the most interesting of the islands are Champion, Floreana, and Gardner. Champion in particular has an array of extremely large coral heads unknown in the northern islands. A nearby sunken crater formation known as the Devil's Crown is perforated around its base with tunnels through which a skillful snorkeler can enter or leave—an incredible underwater playground, pool, and enchanted garden. In the deeper water off Champion I photographed an unusual butterflyfish I called the "baby bull" for its habit of flaring its dorsal spines, then lowering its head to threaten me with the array (Illustration 21). At Española (Hood Island), the black cliffs of the coast are battered by huge waves, resulting in several large blowholes. These spumes of water, the energy of the waves funneled skyward by their own momentum, sometimes reach a height of 20 to 30 meters. Along the rim of Hood's high cliffs lives the world's only known colony of breeding waved albatross. There are some 12,000 pairs of these amazing birds. Ungainly and waddling on land, they are great flyers, launching themselves from the high cliffs and soaring out over the water in a sudden transition to grace. Swooping back and forth above the wind-whipped cliffs, these aerial monarchs with their four-meter-long wingspan ride the long ways of the sky.

One unforgettable incident occurred on the barren rocky beach of Barrington Island. Along the beach were strewn the sleeping forms of perhaps 200 sea lions. A few of them bolted excitedly into the water to greet us. As we landed they watched us curiously, seemingly uncertain whether to flee or play. At one end of the beach a mother sea lion gave birth to a stillborn pup. In obvious distress, overheated in the sun, her body covered with flies, she began to nudge the tiny corpse. When it did not respond, she picked it up in her mouth and carried it away from the wave-wash. Then she began uttering sounds that were very like sobs. Time and again she nudged her dead baby, crooned her distress, and carried her tiny burden from one place to another. Other sea lions from the herd came to investigate, but she chased them away.

Then, high in the sky appeared a speck which resolved itself into an approaching Galápagos hawk. This graceful predator sailed slowly down the wind until it was looking directly down on the drama from perhaps five meters above. Hovering there almost without motion, its sharp eyes took in everything. Slowly and deliberately, the hawk settled down on a rock perhaps a dozen meters from the anguished mother. Frantically she licked her little corpse; then, when a wave splashed it, carried it further onto the rocks. From her body spilled the afterbirth, standing out stark and red against the rock and sand. Within minutes a second, then a third and a fourth hawk settled on the rocks. It was time for us to leave, but we did not want to abandon the mother. The hawks moved closer, until one sat just above the mother's head. When we left, the ring was closing, the drama nearing its inevitable denouement. The mother sea lion would bear another pup, but her apparent grief touched us deeply. Somehow she became a symbol of how close man can approach to nature's most intimate moments.

I have ended every trip to these remote islands reluctantly. There are so many diving sites, so many beautiful encounters that I leave with the feeling of having been graced with a moment of enchantment. Residents of the islands tell me this sense of awe is merely a touch of Galápagos fever, carried by many humbled visitors back to the bustling, troubled modern world.

183. *A brightly colored, slate-pencil sea urchin* (Heterocentrotus mammillatus). *It can remain exposed during the day because of its strong protective spines.* (*Hawaii*)

184. *A yellow-tailed surgeonfish* (Prionurus punctatus). (*Baja California*)

185. Urolophus halleri, *a common ray, in Baja California waters.*

186. *The guitarfish or shovel-nosed shark* (Rhinobatus productus). (*Baja California*)

187. *The giant hawkfish* (Cirrhitus rivulatus). (*Baja California*)

188. *A golden seahorse* (Hippocampus ingens) *taking refuge amid the arms of a gorgonian.* (*Baja California*)

189. *A ribbon of eggs, deposited by a nudibranch, or shell-less snail, on a lava boulder.* (*Kona Coast, Hawaii*)

190. *A pair of painted prawns* (Hymenocera picta) *preying on a Hawaiian starfish. Many photographers regard them as the most beautiful shrimp in the sea.* (*Maui, Hawaii*)

191. *The small boxfish* (Ostracion meleagris). (*Kona Coast, Hawaii*)

192. *A shy spiked shrimp* (Saron marmoratus) *near its coral refuge.* (*Maui, Hawaii*)

193. Acanthurus achilles, *the distinctive achilles tang of Hawaii's reefs.* (*Kona Coast*)

185

186

187

188

190

191

192

193

194. *This yellow sea urchin is common in the shallow waters of the Galápagos Islands.* (*Bartholomew Island*)

195. *A large starfish of the deeper Galápagos waters.* (*Cousins Rocks*)

196. *A shallow-water bat starfish.* (*Bartholomew, Galápagos Islands*)

197. *A bat starfish.* (*Guy Fawkes Rocks, Galápagos Islands*)

198. *A bat starfish found in shallower waters.* (*Bartholomew, Galápagos Islands*)

199. *Brilliantly colored algae and gorgonians decorating stark lava boulders in a Galápagos cave. The reason for such color in utter darkness is not known.*

200. *The many-armed starfish* (Heliaster multiradiata). (*Tagus Cove, Galápagos Islands*)

201. *Club-spined sea urchins, coated with encrusting sponges and algae. This phenomenon is seen frequently in the Galápagos.*

202. *A brightly-hued gorgonian sea fan growing on a lava boulder.* (*Galápagos Islands*)

203. *A delicate anemone blooming on a deep slope more than 35 meters deep near Tagus Cove on Isabela Island.* (*Galápagos Islands*)

204–207. *Galápagos marine iguanas* (Amblyrhynchus cristatus) *feeding on algae which grow on submerged boulders of lava. The iguanas seem to feed only at low tide. They swim to offshore sites, dive down to coral boulders, hang on with their claws, and tear off the algae with their teeth.* (*Punta Espinosa*)

208. *The Galápagos hogfish* (Bodianus diplotaenia) (*Punta Espinosa*)

209. *A horned moray eel emerging from a vaulted undersea cavern.* (*Cousins Rocks, Galápagos Islands*)

210. *A Galápagos slipper lobster.* (*Tagus Cove*)

211. *The almaco amberjack* (Seriola rivolana) *in open water.* (*Gordon Rocks, Galápagos Islands*)

212. *The Galápagos scorpionfish* (Scorpaena mystes). *Its adornments are very effective against a background of algae-covered lava boulder surface.* (*Daphne Island*)

213. *The Galápagos horned shark* (Heterodontus quoyi.) (*Punta Roca Vicente*)

214. *The long-spined sea urchin* Diadema. *It is a hazard for divers, since its sharp spines penetrate gloves or wet-suit easily.* (*Champion, Galápagos Islands*)

195

196

199

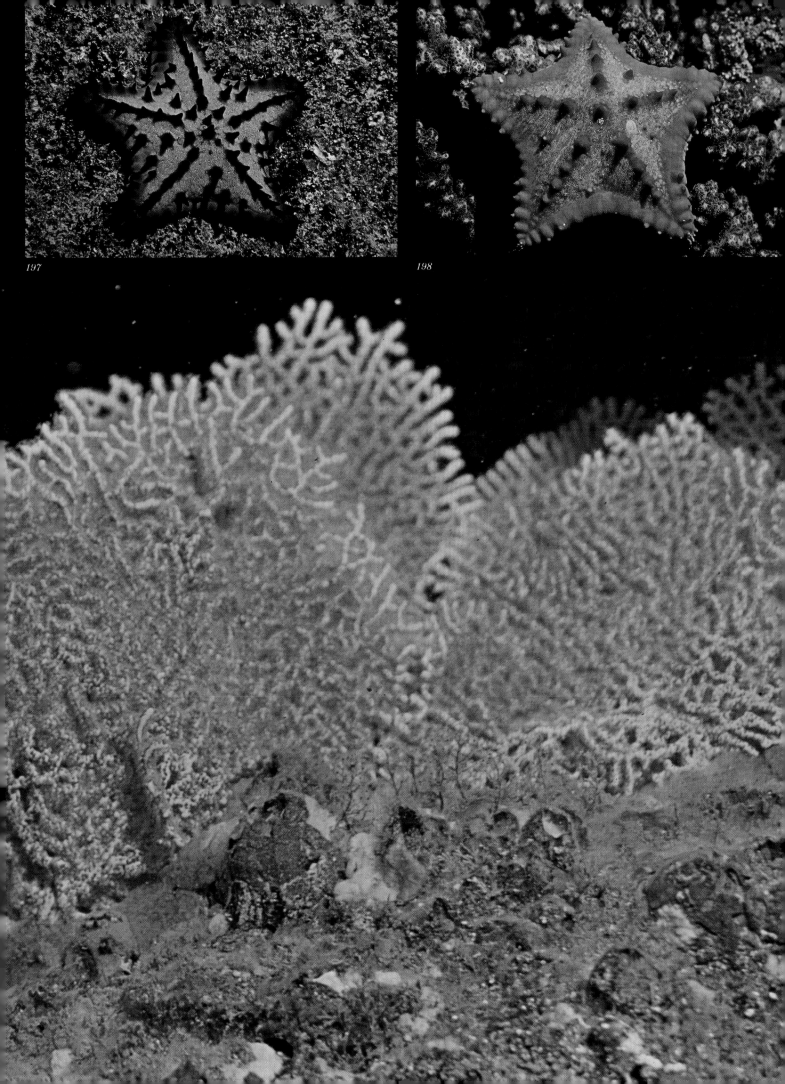

197

198

201

202

204

205

208

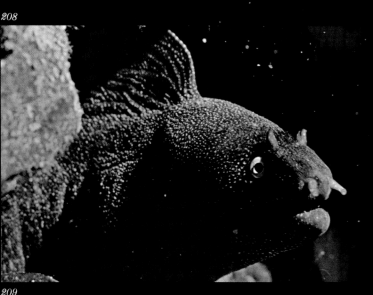

209

210

206

207

211

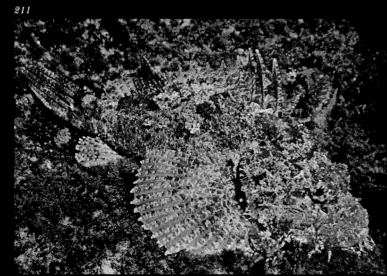

212

213

The Tropical South Pacific

In our culture "South Sea Island" is an overworked cliché, a place of legends, with sultry moonlit nights, swaying palms, white-sand beaches, and a sparkling sea. It summons up visions of *Mutiny on the Bounty*, Paul Gauguin, "His Majesty" O'Keefe, and the cargo cults. For some there are memories of valor: Bougainville, Peleliu, Guadalcanal, Iwo Jima, Saipan. The Pacific is all of these images, and a multitude more. This romantic and dramatic setting is the background of our search for the Pacific's underwater wildernesses.

The reality lies at the end of a seemingly interminable airplane journey. The long flight provides a first insight into the immense distances of the Pacific. After one has been in a jet aircraft for 15 hours over open ocean, one begins to appreciate the exploits of the early mariners. In wooden canoes or tiny sailing vessels they navigated these vast distances by reading the sky, the wind, and the sea. Their skill and daring supported them in epic months-long voyages that we complete in a few hours.

French Polynesia

The islands of French Polynesia are perhaps closest of all to the ideal of our cliché. The Society Islands, of which Tahiti—the very name sounds a siren call—is the largest, are among the most beautiful in all the Pacific. Volcanoes, softened by verdant greenery, thrust upward from calm lagoons. The lagoons are protected from the blue sea by broad reef shallows. Tahiti was discovered for Europeans in 1767 by the English captain Samuel Wallis, who claimed it for his country. A year later the French captain Louis de Bougainville arrived in Tahiti and compared it to the Garden of Eden. Unaware of Wallis' visit, de Bougainville claimed it for France. In 1770 James Cook, then a young lieutenant aboard *H.M.S. Endeavour*, observed the transit of Venus across the face of the sun from an observation fort on Tahiti. Cook charted many of the surrounding islands, naming them the Society Islands for their closeness to one another. It remained for one of Cook's early shipmates, Lieutenant (later Captain) William Bligh, to join the name Tahiti to mutiny and drama.

Rather than follow Bligh's harsh orders, the bulk of his crew, led by First Officer Fletcher Christian, mutinied. They set Bligh and 18 of his supporters out on the sea in an open whaleboat. Bligh managed to navigate this small craft nearly 5,000 kilometers, reaching Indonesia. The British Navy scoured the Pacific for the mutineers; 14 were captured on Tahiti in 1791. The incident was immortalized in the novel *Mutiny on the Bounty* (1932) by Charles Nordhoff and J. N. Hall.

Today the islands are as beautiful in most ways as when Wallis and de Bougainville first saw them. The lagoons are peaceful, the jagged peaks softened with lush jungle green. On the outer reefs, the visitor finds clear water, coral growth which is diverse though not large in scale, and a rich fish fauna.

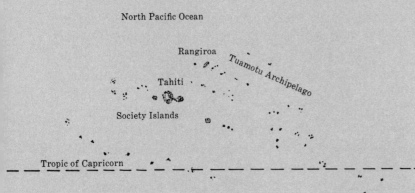

North Pacific Ocean

Rangiroa

Tuamotu Archipelago

Tahiti

Society Islands

Tropic of Capricorn

South Pacific Ocean

The islands of French Polynesia range from high peaks surrounded by quiet lagoons in the Society Islands to vast atolls in the Tuamotu group.

Among the most enchanting of these islands are the Leewards, which lie some 160 kilometers northwest of Tahiti. Part of their special charm is that they are small and therefore can be seen in their entirety even from nearby. Sailing among these jewel-like islands, one always has the next one in view.

The richest reefs in the islands are found outside the passes, or cuts, in the surrounding reef through which swift tides course in two daily cycles. In these passes, where the tidal waters reach speeds of several knots, coral growth is suppressed by current-borne lagoon sand that smothers the polyps. Large populations of pelagic species such as jacks, barracuda, and sharks are found there, preying upon smaller fish that feed in the food stream created by the swift waters.

Outside the passes, away from the smothering sand, a variety of coral species abound. Few of these colonies are very large. They generally consist of a sloping outer reef face covered with coral structures less than one meter in height. By contrast, similarly situated corals of the Tuamotu atoll group (80 atolls some 500 kilometers east of Tahiti) reach heights of two to three meters or more.

In the inner lagoons of the Windwards and Leewards, including such islands as Tahiti, Bora Bora, and Tahaa, grow both shallow-water dome corals and trees of intricate branching corals that reach heights of three to four meters. The daily tidal water movement provides nutrients for these fast-growing colonies, while the protected lagoon waters shelter their vulnerable branches.

The fishes of French Polynesia are highly variegated and rich in both numbers and coloration. The butterflyfish are graceful bursts of color amid the sober hues of the stony corals; the long-snouted butterflyfish (*Forcipiger flavissimus*), the lemon butterflyfish (*Chaetodon miliaris*), and the masked butterflyfish (*Chaetodon lunula*) seem particularly abundant. Other noteworthy species are the hawkfish (including one canary-yellow species), imperial angelfish, flounder, tangs, surgeonfish, lionfish, and groupers. While photographing butterflyfish one day on the outer reef of Rangiroa Atoll, I looked up and saw a manta ray measuring about five meters in wingspan, with a full complement of remoras under its wings. As our dive boat circled noisily overhead, the majestic giant banked slowly and gracefully out over the reef edge into deep water and disappeared. There is an aura of gentleness about these remarkable plankton-feeders that communicates itself to the human observer; one's reaction is to reach out, to try to touch these great undersea flyers. On a peaceful reef, without noisy boats and too many divers, they have frequently allowed themselves to be ridden by humans, or have played about the visitors for hours.

One brightly colored and prolific inhabitant of the shallow, still waters just inside the barrier reefs is the Tridacna clam. Resplendent in the iridescent

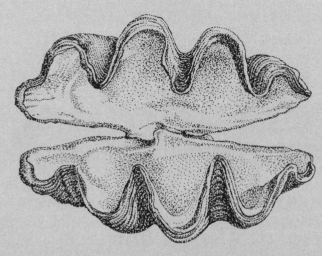

Shell of a Tridacna clam. This mollusk can reach a size of 1.5 meters and weigh more than one hundred kilograms.

mottled blue, brown, or green of their symbiotic algae, these mollusks sometimes occur in clusters of a dozen or more on one small coral head. Frequently side-by-side with the Tridacnas are the serpulid worms of the genus *Spirobranchus*. In this region of the Pacific they occur in bright yellows, burgundy-and-white combinations, and occasionally even in blue (Illustration 259).

Quite different denizens of these calm lagoons are several species of moray eels. Though principally nocturnal feeders, the eels show no hesitation in extending nearly half their two-meter-long bodies out of a hiding place to take food from divers, never once threatening the generous fingers. Smaller morays, a meter long or less, may even take proffered fish which cannot fit through the entry of their shelter. After several attempts to shift the prey in its grasp, the moray seems to lose patience. Then in a flurry of scales and stirred-up sand, it and its mauled meal disappear into the coral sanctuary.

Among my most striking encounters in the Polynesian islands were those with the octopus. On three occasions I met fairly large octopuses whose curiosity seemed to outweigh their caution. In each episode, the octopus moved rather openly across several coral structures, not even bothering to hide. When it was clear it had been seen, this animal, known among scientists for its intelligent behavior, simply molded its fluid body to resemble a knobby protrusion from the coral. In addition, it suddenly flushed itself a nubbly brown and white. Immobile, blending into its coral head, the octopus impudently watched my camera approach to within a meter or less of its perch. The animal never seemed to fear me, but merely combined its curiosity with a certain caution.

In one case two ham actors stumbled onto the same stage. My wife Jessica and I were filming a well-behaved octopus, when from a crevice beneath it a grouper emerged. Staring at the camera, the grouper insinuated himself into the scene without coming between the octopus and the camera. Throughout our 20-minute visit with the octopus, the curious grouper remained, a discreet observer of the entire scene. Yet never once did it intrude upon the performance of the octopus.

Another gaudy species found frequently on these reefs is the lionfish (*Pterois*). Several species occur on the islands, the most richly decorated being *Pterois antennata*. Plumed lances above its eyes add a final touch to its venom-spined finery, a potent defense. While most often encountered singly, the lionfish can sometimes be found here in clusters of six or more sharing a coral shelter. In these close quarters the lionfish particularly seem to resent intruders.

A certain excitement on the reefs and especially the passes of Polynesia are provided by sharks. Rangiroa, with a length of 65 kilometers, the largest of the Tuamotu atolls, has but two passes, Avatoru and Tiputa. Twice each day a large volume of tidal water

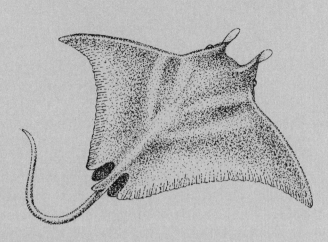

A manta ray, one of the largest of living fishes, may have a wingspan of up to eight meters.

must move in and out of the vast lagoon through these passages, each less than one quarter of a kilometer wide. In the Tiputa pass, the inrushing current sweeps divers past a kilometer or more of reef to a dark valley. Here, in a hushed atmosphere, large coral heads tower above bare sand in suddenly diminished light. Patrolling the valley is a substantial population of white-tipped reef sharks (*Triaenodon obesus*). These one-to-two-meter sharks seem relatively peaceful. However, the simple act of spearing one or two fish alters the situation perceptibly. The vibrations transmitted by the thrashing of a wounded fish instantly draw one after another of the suddenly aggressive sharks. Moving in twos and threes now, they seek the source of the distress vibrations. While not dangerously fast-moving, the sharks sweep close to me as I sit on a small coral stool which is fine as a photographic vantage point but offers no protection whatsoever. I realize that the sharks may soon mistake me for the bait. Moments later I am fending off the sharks with my camera while my diving companions delightedly film my predicament, thinking it somehow prearranged.

A few seconds later an entirely new variable enters the equation. From their shelter within the large coral heads, a pair of green moray eels are hotly contesting the wounded fish with the shark pack. Sharks dive head-on into the coral, but they are not accustomed to close-in maneuvering among these complex obstacles. Within moments the highly maneuverable eels have snatched the prey from the jaws of the sharks and swallowed it.

Now, into the middle of this turmoil drifts a manta ray, a small specimen with a two-meter wingspan. Slowly, slowly, wings motionless, the manta allows its momentum to carry it into the very center of our arena. While we remain motionless, the ray edges right through the cluster of divers, sharks, eels, and coral heads. Once safely beyond us, the inoffensive creature slowly moves its wings again and soars off into the gloom of the valley.

One unusual fish of the Polynesian reefs is silver-blue and shaped like a damselfish, with royal-blue horizontal stripes on its lower sides. In fact, it is *Genicanthus watanabei*, an angelfish which in its evolution has found advantage in both resembling and behaving like a damselfish (Illustration 217).

On another day on Rangiroa's outer reef I encountered an orange-yellow pufferfish (*Arothron nigropunctatus*). Having often enjoyed filming these rather puppy-like fish, I moved closer and discovered that the puffer exhibited no fear at my approach (Illustration 218). Gingerly I moved in until I could reach out and touch it. For several minutes the puffer posed, then seeming to tire of the sport, swam in a galumphing way down the reef beyond my reach. While the low-growing corals of the outer reef slopes in the Society Islands (such as Huahine, Raiatea, Tahaa, and Bora Bora) would not be considered outstanding by divers, they, like the Hawaiian Islands, are endowed with a diverse and plentiful fish fauna. Rangiroa and the other Tuamotu atolls, on the other hand, possessing the same fish species and massive coral colonies, would have to be considered in any ranking of "best" underwater wilderness sites. The beauty of these islands extends from their volcanic peaks to the rich marine life on their undersea flanks.

Micronesia

While French Polynesia is composed of 130 islands and atolls, Micronesia has 2,141 islands scattered over nearly five million square kilometers of the Pacific. These islands, also known as the Trust Territory of the Pacific Islands for their trusteeship status, have a long and tumultuous history.

In the centuries before European colonization, the six major island chains (politically organized into the Marshalls, Ponape, Truk, Marianas, Yap, and Palau districts) were ruled by tribal chiefs. Some of the tribes numbered from 50,000 to 100,000 members. Many of the people were not only sophisticated navigators, fishermen, and ocean-ranging traders but also highly skilled artisans. Indeed, in the Ponape district are found extensive stone cities built on an immense scale. The best known of these, Nan Madol, on Ponape, had temples encompassing more than 100 city blocks; the ruin covers more than 15 square kilometers. The temples are built of polygonic stones resembling huge logs. Many of the stones weigh several tons each, and were quarried more than 20 kilometers from the site of Nan Madol. The temple walls of the main structure, some of which are three meters thick and ten meters in height, stand silent today, partly overgrown with jungle. The civilizations that built these monumental cities would hardly have been mere local fishermen.

In their pre-European history, the tribes in the various regions evolved quite different societies. Only in Yap, for example, did the natives undertake hazardous ocean voyages to quarry stone for donut-like discs. These stones, up to four meters across, were used as money. Only in the Marshalls did native navigators devise tide and wave maps made of sticks and shells. Only in Palau were the homes decorated with intricate bas-relief carvings.

With the coming of European explorers and missionaries, however, everything was altered. The first tribe to encounter the Europeans was that of the Chamorros of the Marianas Islands. The Chamorros, a tall, fair-skinned people of great attainments in navigation and trade, had a stratified and stable social order. They were first discovered by Ferdinand Magellan in 1521 during the Spanish explorer's circumnavigation of the globe. From the mid-1500s to 1668 the Chamorros enjoyed occasional but not dangerous contact with Spanish ships that were plying their regular trade routes to the Philippines. With the eventual settling of the missionaries, however,

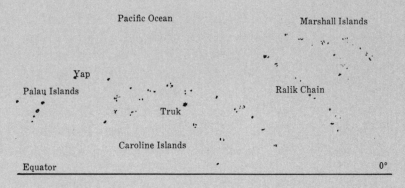

Consisting of more than 2,100 islands scattered over millions of square kilometers of ocean, Micronesia sweeps across the central and western Pacific.

came social upheaval. In 1672, only four years after the first missionary, Father Sanvitores, arrived, he was murdered for baptizing a baby without its parents' permission. Sporadic war followed for 25 years, with the Chamorros repeatedly decimated in combat, by internment, and by exposure to European diseases. By 1710, only 3,500 Chamorros remained. These embittered survivors, unable to cope with the changes demanded by the Europeans, took up abortion and infanticide and in the end destroyed their own race.

Similar exterminations of indigenous populations by disease and conflict occurred throughout Micronesia during the 18th century, and much of the original populations disappeared. For three centuries, one colonial power after another annexed the islands, realizing their strategic mid-Pacific importance. Spain held the islands until 1885, when the Kaiser's Germany achieved virtual domination.

It was also in 1885 that one of the Pacific's most colorful characters, David ("His Majesty") O'Keefe, a swaggering Irishman, emerged to play his role on the Micronesian stage. O'Keefe discovered that with modern tools and ships he could obtain from Palau the stone discs the Yapese used for money. For the Yapese, the 400-kilometer transport of the huge stones had been fraught with hazard; for O'Keefe it was easy. He used his quarried money to purchase bêche-de-mer (sea cucumber, an oriental delicacy), trochus shell, and copra from the Yapese, and invested the proceeds in Hong Kong real estate. Local resentment in Palau against O'Keefe's activities started a conflict in which one of his ships was burned. The British intervened on O'Keefe's behalf. The Spanish, fearing not only the Germans but now also the British interest in Micronesia, pressed their case and finally won a settlement in which the Pope, as arbitrator, ruled for Spain.

In 1898, however, the Spanish-American War effectively removed the Spanish from Micronesia, leaving the islands to the Germans. Then World War I ended German hegemony and the islands were invaded by the Japanese. After the war, the Americans, British, and French challenged the Japanese occupation. The Japanese, with their own global designs in mind, offered numerous guarantees and were awarded custodial authority over the islands.

The Japanese were the first conquerors to colonize the islands. By 1942, some 80,000 resident Japanese were the key to Micronesia's entire economy. Their exit in the thunder of World War II wracked Micronesia socially and economically as well as physically. Allied military assaults upon the Micronesian islands were savage and bloody, the islands were shelled and bombed, and their harbors were littered with sunken ships.

Since the war, American policy has been one of assistance but not involvement, financial aid but not colonization. Because of this policy, the economy of Micronesia has never really recovered, and some of the inhabitants remember fondly the infrastructure of local trade and commerce under Japanese rule.

Truk   Nowhere is the savagery of Micronesia's recent history more evident than in the lagoon of Truk Atoll. The lagoon is nearly 60 kilometers wide, with scattered islands and a protective outer reef. Here in 1944 was located the main base of the Japanese Imperial Fourth Fleet. In mid-February of 1944 the American Navy launched a major attack. After two days of repeated assault, some 64 ships, 250 aircraft, and thousands of bodies littered the lagoon. The stench of decomposition lingered for months, and even today in the bright sunlight there is something dark and bloody at the edge of the mind as one looks out over the calm lagoon.

In the past few years Truk has achieved a great reputation among underwater enthusiasts who have dived on the eerie naval behemoths beneath its placid waters. In one venture a Japanese-American expedition recovered the bones and personal artifacts of 84 crewmen lost on the submarine *Shinohara* (I-169). The filmed story was subsequently seen on television in the United States.

Some shallow wrecks in Truk, most particularly the *Shinkoku Maru* and the *Fujikawa Maru* (whose masts protrude from the water), are festooned with incredibly prolific coral growth. The *Shinkoku Maru* is a tanker sitting upright on the bottom of the lagoon. Her masts are intact and only five meters beneath the surface. Her railings, masts, smokestack, and gun are ablaze with sponges, soft corals, and hard corals and filled with swarms of both small reef fishes and pelagic species. Indeed, the *Shinkoku* is one of the richest coral reefs I have ever seen.

The *Fujikawa Maru* (or *Fuji*, as she is familiarly known) is a freighter 150 meters in length which sits upright with her deck a mere 10 to 15 meters below the surface. Both her fore and aft masts protrude from the surface. In the warm, sunlit waters of the lagoon the coral growth on the exterior of the *Fuji* has been extraordinary. Scalloped coral colonies (*Sarcophyton*) grow to more than a meter across. Her bow as well as her stern cannon are adorned with large colonies of branching corals that almost completely disguise the weapons' original lethal shape. Great concentrations of gorgonians frame what once were doorways. Above all, there is one of the world's greatest aggregations of soft corals. In shimmering yellow, rose, pink, and a hundred other hues they hang from the cross-members of masts, railings, and cables. All around this wreck has assembled the remainder of a full reef-life spectrum, including clouds of damselfish, butterflyfish, anchovies, and other dainty reef fish.

Other wrecks, also in fairly shallow water, have a less plentiful coral growth, possibly because they are not in the direct path of coral-larvae-bearing and plankton-rich currents from the outer passes. Among these are the *Rio de Janeiro Maru*, the *Kiyozumi Maru*, the *Hoyo Maru*, the *Heian Maru*, and the *Yamagiri Maru*.

There are also fairly deep wrecks which have acquired living adornment over the years. Among these the *Seiko Maru*, some 40 meters below the surface, is known for its lavish encrustation of sponges, finger-like gorgonian colonies, and crinoids. Finally, there are the very deep wrecks such as the *San Francisco Maru*, lying 60 to 80 meters down. These are truly the ghost wrecks of the fleet, beyond even the blessing of sunlight.

The wrecks of Truk Lagoon are a moving testament to the horror of war. Their death-still cabins and holds contain both lethal and pathetic reminders. In the holds of the *Rio de Janeiro Maru* are huge cannon barrels. The *Yamagiri Maru's* central hold is a tumbled mass of projectiles, used for the 18-inch guns of Japan's mightiest battleships. On the *Shinkoku*, by contrast, we found phonograph records, uniforms, and other poignant personal effects of the men who were killed with their ships. Many of the ships were wracked by giant explosions from bombs and torpedoes; the blast which destroyed the *Aikoku Maru* blew an American fighter plane right out of the sky. Similarly, the capsized hull of the *Hoyo Maru*, with an enormous hole in it, is even visible from aircraft flying overhead.

These sunken vessels are a magnificent laboratory in which to study marine life growth under a variety of identifiable conditions. In addition, they are one of the most awesome man-made additions to the underwater wilderness, a dazzling—and sobering—spectacle.

Palau

The other major underwater wilderness area of Micronesia is Koror (Palau). Palau's history is as ridden with turmoil as that of the other districts. Now one sees only sun-drenched peace in her rock-garden islands. When seen from the air, more than 200 of these abrupt islands seem to be scattered in the water like green mushrooms. At water level one discovers that they are coral structures rounded from long erosion. In the process of breaking down, the limestone was transformed into a covering of nutrient "soil" in which hardy greenery has taken abundant root. Most interesting is the fact that all of these islands have been strongly undercut at the water-surface level by wave erosion and the feeding of a shellfish known as the iron-tooth chiton.

Off several islands such as Babelthuap and Kayangel Atoll, rich groupings of marine life exist due to food-bearing currents that concentrate nutrients. One excellent site is at Ngemelis Pass, a deep slash through Palau's outer reef perimeter. When the tide is rising, clear ocean water races inward through the pass, enabling the visitor to photograph the prolific gorgonians, crinoids, and soft corals that line the plummeting wall of the pass. At the top of the wall, a coral garden basks in knee-deep water filled with tiny reef fish that find relative safety from larger predators in such shallows.

Other areas of Micronesia, including Guam and Saipan (both belonging to the Marianas Islands) and Ponape and Majuro (Marshall Islands), are now being opened to divers. Some of these islands will prove to be as rich in particular species as those of Truk and Palau. But if we recall our earlier chapter on the Caribbean, Cozumel has a structure, a uniqueness, which gives it a special place in undersea exploration. The same is true for Truk, with its ghostly fleet, its generation of painful memories slowly echoing into the past. Palau, too, with its maze of randomly shaped garden islands and its steep undersea walls swarming with life, has defined its place under the Pacific sun.

New Caledonia

The sight of this group of mountainous islands recalled Scotland to Captain James Cook when he first saw them in 1774, and he gave them a version of Scotland's ancient name. The main island of the group, often called Grand Terre, is divided in two by mountains running its entire length. The western side of the island is relatively flat and densely populated with a white-barked gum tree called *niaouli*. More than half of New Caledonia is covered with crystalline serpentine rock, and its mountains are scarred from major mining operations that extract nickel, cobalt, iron, manganese, and chrome. New Caledonia has a large coral reef surrounding it, and is one of many locales claiming "the second largest barrier reef in the world."

Like so many other islands of the Pacific, New Caledonia was a pawn in the global struggles of the European maritime powers in the 17th century. In a close race between Britain and France to annex the island, the latter won out in 1653. Justifying the annexation on the grounds that some of their citizens had been killed on the island, the French installed one of the most repressive and brutal penal colonies in the Pacific. After 1897 the French halted the movement of convicts to New Caledonia, but inevitably prisoners and former prisoners made up a large proportion of the French population there. The native Melanesian population declined drastically after only a century of French domination, from a peak of 60,000 to 40,000. Diseases, which swept the entire Pacific, also took a large toll of these ancient seafaring peoples.

About 50 kilometers southeast of New Caledonia lies Kunie, better known as the Isle of Pines. The island is beautifully forested with arrow-straight pines which reach a height of 50 meters. Now an endemic species, these trees are descended from original stock imported from Australia. The Isle of Pines is known for two spectacular underwater wilderness sites, the Gadji Pass and the Grotto. The Gadji Pass is a tidal cut through the outer reefs that surround Kunie.

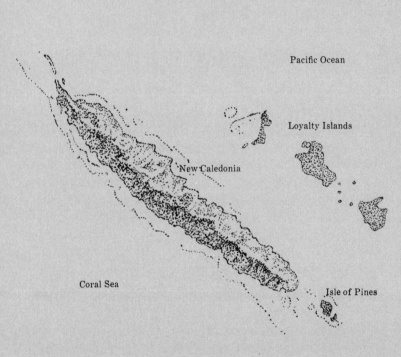

Pacific Ocean

Loyalty Islands

New Caledonia

Coral Sea

Isle of Pines

Renowned for its nickel mines and exquisite reefs, New Caledonia is among the most richly endowed of the South Pacific islands.

Along the edges of this huge pass are thick walls of coral extending from a three-meter depth to 100 meters or more. In the shallows atop the wall a carpet of scalloped corals, hard corals of various species, and brilliant red bushes of the gorgonian *Mopsia ellisi* completely cover the coral surface. Strong currents sweep over these shallows, providing nutrients for the lush growth of coral colonies.

At several points along the great coral wall one finds enormous canyons, cracks in the reefs, probably caused by earthquake activity. (New Caledonia is in an extremely active tectonic zone.) Since the cracks lie perpendicular to the current flowing through the pass, the diver can follow an anchor line down to the reef-top, then escape the swift water flow in the pass by dropping into one of these canyons. The width of the cracks ranges up to 10 meters, and they often display overhangs and smaller radiating cracks. The walls are sheer limestone rock, coated with varicolored coral polyp clusters, brilliant orange-red gorgonians, and intensely colorful members of the soft coral family. Schools of damselfish, surgeonfish, grunts, and parrotfish swirl in and out of the jagged open skyline above as you hover in the relative stillness of the canyon. Butterflyfish and emperor angelfish sweep past along the canyon walls, cunningly using small crevices to evade your reach. In the open deep blue of the pass, seen beyond the canyon mouth, patrolling jacks, large groupers, and sharks are shadowy silhouettes in the fast-moving water.

Across the pass is a very different reef configuration. Small valleys are cut in the reef-top, and large gorgonians fan themselves into the channels thus formed. Then, abruptly, the coral drops away and you are looking into deep water. There, huge schools of jacks, surgeonfish, and groupers mill in the racing waters just off the steep face of the reef, while sharks restlessly patrol in deeper water. The face of the wall has soft coral trees in canary yellow standing nearly a meter in height (Illustration 215), and orange gorgonians that are even larger. As soon as the noisy divers arrive, the schooling fish move on, leaving only the resident reef species and the sharks.

In one area about a kilometer farther along the wall is a quiet lagoon zone with an almost square reef-top nearly 30 meters deep, surrounded by sheer plunging walls. This reef is covered with low-growing corals and anemones, while along its edges schools of moorish idols, damselfish, angelfish, and fusiliers dance and swirl in the eddying currents.

A few kilometers west of Gadji, in shallow water, one of my diving companions and I came upon a remarkable scene. As the two of us sat motionless for over an hour, growing cold from inactivity, a pair of maroon flatworms performed what could only have been an elaborate courtship ritual. At first the flatworms would move along the algae-covered coral rock surface on convergent courses until they nearly touched. Then the larger of the two reared up with the for-

ward half of its body like some tiny king cobra. As we watched transfixed, the upright worm thrust from the underside of its body two protuberances resembling breasts and extended them searchingly toward its companion. The second worm reared and extended its own organs. Groping toward each other in a way that seemed almost human, the two worms finally touched. At the touch, both worms retreated and lowered themselves to their normal flat orientation. Again and again during the hour we watched them, the two worms reared, touched, and retreated, only to repeat the encounter a few moments later. Shortly before we ran out of air and were forced to surface, the two worms finally seemed to complete their ritual. Abandoning their daintily majestic upright stance and resuming their common flat appearance, they slowly moved across the algae-carpeted rock on their separate paths.

For another remarkable experience on the Isle of Pines, we had to leave the reef and journey inland to the Grotto, an unusual underground formation of rare beauty. Parking our car in a thick woods, we dressed completely for a dive. It seemed somewhat odd to put on our heavy diving equipment and walk through the woods when the ocean was nowhere in sight. Beneath the gnarled roots of several tall trees we entered a gloomy cave entrance and climbed down a twisting dirt path punctuated with large stalagmites. The limestone cavern was as large as a church, and pitch dark; far off in its echoing stillness we heard a trickle of water. After climbing downward perhaps 35 meters, our French dive guide stopped us. We turned on our night lights, and saw that mirror-calm water began at our feet.

We slipped into the cold, clear water. Swimming a few meters, we went single file into a small crevice and entered open water of utter blackness. We had been warned to control our buoyancy in this water-filled cavern, and flashing my light downward I understood why. A thick coating of fine green silt covers the floor of the grotto. In many places the water reaches the roof of the cavern. If we stirred up this silt it would cloud the water and prevent us from finding our way out. This is clearly a dive for the experienced diver, as the penalty for error is very high.

Above us our darkness-cloaked bubbles crashed into the roof of the cave. Where there was no open surface above us we were highly dependent on our diving equipment.

The grotto was formed by the dissolution of its limestone, perhaps by an underground river. Then during an ice age when the sea level was greatly lowered, water seepage from the forest floor above formed the profuse stalagmites and stalactites by depositing limestone salts. The end of the ice age raised the surrounding sea level once more, and the grotto was flooded. Now, as we tried to pierce the darkness with our lights, great stone shapes loomed suddenly in our path. We found what seemed like walls of organ pipes but are actually a maze of small stalactites. Off the main grotto, dozens of side passages wait to trap the unwary. We followed our host's directions for the dive precisely. It does not pay to be bold here. After a while we wrenched ourselves from the eerie spell of this chamber, and returned to the outer cavern.

As we left the island the next day, our plane circled first over the forest, then over the magnificent reefs of Gadji. Somehow the two environments would never again seem as distant from each other.

The New Hebrides   These enchanting islands, one hour northeast of New Caledonia by jet, form an incomplete double chain stretching north and south for nearly 700 kilometers. The islands are volcanic in origin and geologically relatively young. There are five active volcanoes in the group.

The first Melanesians arrived in the New Hebrides about 1000 B.C. They sailed from the Papua–New Guinea area, spread throughout the Solomons, and eventually reached the New Hebrides. Somewhat later, Polynesians came to the southern islands, probably from Tonga or the Cook Islands.

Pedro de Quieras, a Portuguese searching for the legendary "Southern Continent" that turned out to be Australia, tried to establish a colony here in 1606. However, the colonizers, defeated by illness and disputes with the natives, soon withdrew. The islands were then ignored until they were visited by de Bougainville and James Cook in the years 1768 and 1774, respectively.

In 1825 an Irish seaman found sandalwood, whose oil was used in perfume, on the islands. Demand for the wood drew felons, pirates, and smugglers to the area. Close on their heels came the missionaries. As we have seen so often now, the combination of greed, zealotry, and disease brought catastrophe to the natives. "Blackbirding"—capturing natives for sale elsewhere as slaves—also resulted in many deaths from starvation, sickness, and ill treatment. In 1886 Britain and France nearly went to war over the islands. But, pressed by German expansionism in the Pacific, they formed a joint or "condominium" government in 1906.

There are excellent reefs and a plentiful fish population around Efate and other islands. In addition, one of the world's most placid and beautiful lagoons is located at Lagon d'Erakor near Port Vila. But for one of the vast wonders of the underwater wilderness we fly to Espiritu Santo, a northern island of the chain. The tropical sun beats down on the orange-clay roads of the sleepy town of Luganville, producing a shimmering dusty haze. What does move, moves slowly, in small cars and old trucks. There is but one little hotel and apparently no tourists.

Each day there are several small earthquakes, just little rumbles. The natives say you can tell when there's a quake, because the few overhead wires on the main street hum and buzz. According to scien-

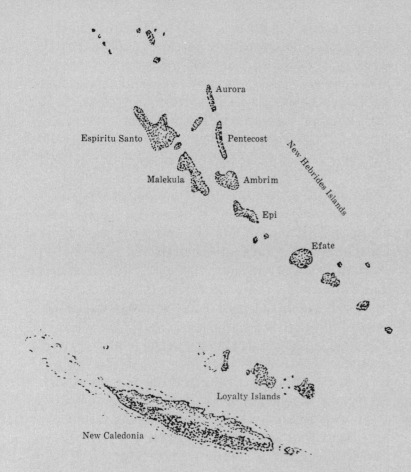

The New Hebrides, described as "islands of coral and ashes," are characterized by active volcanoes and prolific reefs.

tists, more than 2,000 such tremors are recorded here every year. Other than that there's very little excitement, just tranquility and forgetfulness.

The waters off Espiritu Santo are quiet now. It was not always so. At the height of World War II, this tiny island was one of the most strategically important naval bases the Allies held, a staging area for massive military thrusts to end the Japanese hegemony in the South Pacific. Places with bloodily familiar names were not far over the horizon: Guadalcanal, Bougainville, Rabaul, Munda, the Coral Sea. This was a place where men arrived young and left old and tired and hurt, if they left at all. Today, looking out over the blue waters of the main channel, it is almost impossible to imagine the vast armadas that assembled here. The huge military reservations are forgotten and overgrown, the many thousands of men have long since departed.

There are few reminders left in Luganville. A kilometer east of town, off a sandy beach with an emerald-green fringing reef, you can look out over the empty, sun-lit water, where war has left the biggest reminder of all. In the blue waves, streaked with lava pumice blown from the volcano on Ambrym Island, oil steadily bubbles to the surface and drifts in a slow rainbowed smear toward town. The slick points to the undersea grave of a great passenger liner, the S.S. *President Coolidge.* In 1942, after years of serving the luxury cruise market, the *Coolidge* had been pressed into service as a wartime troopship. Her brilliant black and white body had been painted a dull, dead gray, and 4,000 troops tramped the salons which had once been reserved for the revels of the rich and famous. Above the decks once set aside for shuffleboard and idle lounging now glowered elevated turrets and cannon.

In December of that year, the *Coolidge* left Espiritu Santo under sealed orders. Shortly after her secrecy-shrouded departure, military intelligence learned that a Japanese submarine wolfpack was moving into the area. Urgent orders crackled out for an immediate return to the relative safety of the naval base at Espiritu Santo. The *Coolidge* turned and steamed for cover. No one is really sure what happened then. It is known that a harbor pilot came aboard to guide the giant ship safely through the protective minefields around the base. The rest of the truth resides in the files of a U.S. Navy Court of Inquiry.

Mortally wounded by two mine explosions, the heavily listing ship managed to struggle around the promontory that protects Luganville from the sea. She was rammed directly onto the fringing reef near the airport outside of town, where her entire complement of troops and crew swarmed down her sides and boarded small boats or swam to the nearby shore. There the vessel rolled over and slipped beneath the surface.

Diving on wrecks in deep, unfamiliar waters is sometimes a dangerous business. We therefore prepared in subdued excitement for our dives, after taking a small motorboat out to the source of the oil slick. When we were finally suited, we slipped over the side and descended slowly through the dark water.

There are experiences which nearly defy description, and our first view of this engulfed leviathan in the undersea gloom was clearly one of them. We touched down on the starboard side of the ship near the bridge, which is now the part nearest the surface. The water was not clear. For a moment we stood motionless, lost in awe at her immense size. Forward, in the gloom, we could make out a large gun mount, while nearby the massive bridge leaned out over the depths.

The *Coolidge* is overgrown with countless life forms: sponges, branching corals, worms, shrimp, gobies, anemones, and algae cover every centimeter of her skin. In the green darkness of the open water around us several varieties of jacks, barracuda, groupers, spadefish, and batfish hovered. The *Coolidge* has become a vast city of undersea life, rife with color and movement. Slowly, tentatively, we began to explore. In her forward hold lay stacks of munitions, scattered aimlessly but still lethal. Above us, toward the glow of the distant surface, her huge davits clawed futilely toward the light. Her guns projected out over the dark waters.

Looking closely at a silent gun, I saw a solitary fringed nudibranch, a delicate shell-less snail, munching its way along the dead cannon, browsing on its covering of leafy algae (Illustration 240). In the dark water beyond, an eagle ray passed. Our dives were coming to an end. The other divers had returned to our small boat anchored above.

For a little while I was alone on the great lost ship. I wandered to the silent bridge, glided through an open window, and sat motionless for a long while as I looked about. The row of square, empty windows rose in the dim light above me, and I thought for a moment about sharply pressed uniforms and decorative braid and orders quietly given. A grouper drifted by outside the windows. I began to find it hard to breathe, since my scuba tank was nearly empty. Slowly, reluctantly, I left the *Coolidge* and drifted upward. I could see the entire forward third of her now, as she receded into the depths as if forever sinking.

The Philippines

The islands of the Philippines archipelago extend for some 2,000 kilometers along the southeastern rim of Asia. The 7,107 islands possess almost twice the coastline of the United States, though their land area is no larger than Italy. Located in a powerfully restless volcano and earthquake zone, the islands contain ten active volcanoes.

There is a local saying that the original inhabitants walked to the Philippines when the sea level was much lower. Scientists believe that in about 10,000 B.C., Malays from what is now Indonesia first came to the archipelago. A further wave of immigration

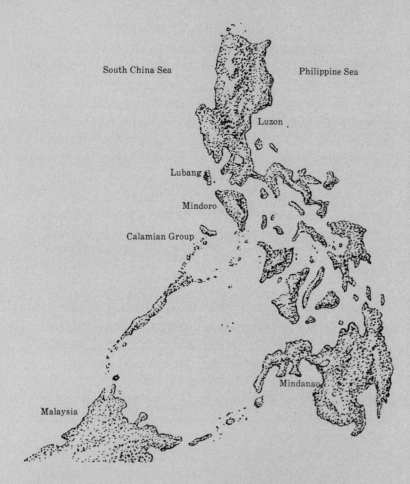

The Philippines, composed of over 7,000 islands with notable extremes of environment, has one of the world's richest marine fauna.

occurred in about 700 B.C., as great numbers of Malays arrived in large family canoes called *barangays*. These settlers brought with them knowledge of rice cultivation and the ability to forge iron tools. Virtually isolated in the islands, the Filipinos retained their ancient ways with very little influence from the outside world until the arrival of Ferdinand Magellan in 1521. Magellan named the islands St. Lazarus. Shortly after his arrival he became involved in a local dispute and was killed. It was 21 years before the Spaniards returned, hoping to share in the rich spice trade that Portugal was developing in Indonesia. They renamed the islands "Las Felipinas" to honor their Prince, the future King Philip II. An army of missionaries soon arrived to convert most Filipinos to Christianity. The Spaniards established Manila on the island of Luzon as the capital and made the Philippines into a major trading center. Silver from Spanish Mexico was traded for Chinese silks, jades, gemstones, and porcelains. The missionaries and the powerful binding influence of the Catholic Church reinforced Spanish control over the islands. By the early 1800s the Church was the largest landowner there. Then Spanish fortunes declined. The Mexican independence movement ended the galleon trade. Independence movements in the Philippines were harshly suppressed but weakened the Spanish grasp, which was abruptly and finally broken by the Spanish-American War. The United States opened vigorous trade with the Philippines. American domination of major agriculture and industry fueled increasingly potent independence movements, but just as independence was scheduled the Japanese invaded the islands. The war was terribly destructive to the Philippines, and both the Allies and the Japanese suffered devastating losses there. MacArthur's withdrawal from Corregidor and Bataan by night and his promise to return are classic events in the annals of war.

Modern Philippine governments have struggled with an uneven economy, social problems, and insurgency. Currently, determined efforts to expand both trade and tourism portend the Philippines' emergence as a major commercial center of the Far East.

The undersea wonders of the Philippines have long been known to scientists. Since the islands lie along the larvae-rich easterly tropical currents, the Philippine marine fauna is considered one of the world's richest. For complex reasons of cost, politics, and distance, these islands have remained virtually untouched by divers other than those serving science and the military. Only the fishing, particularly the dynamite fishing, by the Filipinos themselves, has affected the condition of their reefs. Laws have now been passed outlawing the use of dynamite, and damage to the fauna near population centers should diminish.

Still, cruising across Manila Bay's huge expanse, one encounters literally hundreds of native fishermen in *bancas,* or outrigger canoes. At anchor after dark one sees brilliant points of light scattered across the bay. These are larger fishing boats, each equipped with dozens of thousand-watt lights. The illumination draws fish to the surface, where they are netted. At night in the crowded Manila harbor one cruises past row upon row of fishing boats in a low hum of voices and movement. Many boats have three to five families living aboard. Tiny *bancas* crisscross the channel carrying socializers, their oars splashing gently. All these people and millions more must live off their fishing, so it is no surprise that in some locales overfishing has occurred. Partly because of the pressures of such fishing practices, the finest diving areas in the Philippines are the most remote. Some of the very best underwater wilderness zones lie some 100 kilometers west and southwest of Mindoro Island, in areas such as the uninhabited islands of Busuanga (the Calamian Islands) and the magnificent Apo Reef. In this region the combination of open ocean, steep-sided undersea mountains thrusting to the surface, and highly variable seasonal currents have produced amazingly large populations of pelagic species, reef fish, and colorful invertebrates. Particularly abundant are the broad fronds of umbrella corals, a species prolific throughout the Indo-Pacific as far west as the Red Sea. Large scalloped corals occur in profusion, as do the strongly compacted dome corals. Especially rich are the crinoids, or feather stars, that seem to be a flourishing import from tropical Australia. The coloration of the Philippines species has diverged from that of the Australian. Where the Australian crinoids have a color range of rather bright reds, canary yellows, and greens, Philippine representatives tend toward darker greens or browns, with arm-tips bedecked in gaudy yellow. One crinoid I had never seen elsewhere was a delicate, light powder blue with accents of cream (Illustration 246). Another group of decorative reef-dwellers are the gorgonians, many colonies reaching a span of two to three meters in areas of passing currents.

I have not observed sponges in such large numbers as in the Philippines anywhere else in the Pacific. They occur in great variety and are more reminiscent of the well-developed population of the Cayman Islands in the Caribbean than of other Pacific sites. Several blood-red specimens are found in depths of 20 to 30 meters, while the reef-top shallows have basket sponges over a meter high.

On a remote island reef I happened upon a medium-sized green moray eel, somewhat more than a meter long. Surprised, it raced away at top speed to another coral head, where it turned in a threatening display. The abrupt movement of this predator caused clouds of small reef fish to scatter wildly, and even when the eel settled, a palpable tension lingered on the reef. When I moved in to attempt a closer portrait, the eel withdrew into the reef until only its face and formidable jaws were exposed.

A particularly colorful feature of these remote Philippine reefs was the sizeable schools of triangle butterflyfish which milled in brightly flashing aggregations at depths of 25 to 35 meters. Sometimes a school might number several hundred individuals placidly meandering one or two meters above the coral slope, while gaudy powder-blue fusiliers with yellow backs breezed by on their wider reef rounds. Every coral head here contains some brightly colored fish amid its invertebrate fellows: moorish idols, lionfish, banner butterflyfish, trumpetfish, and many others throng these sculptured meadows. At one coral head my Filipino diving companion discovered a large blue and gold pufferfish in a crevice. We gently urged it out, hoping not to frighten it into flight. After a few minutes of maneuvering, it hovered near its shelter long enough for me to manage some photographs. Eventually it either became frightened or lost interest; it simply swam away rapidly in that ungainly top-to-bottom waddling movement characteristic of the larger pufferfish.

On a nearby reef I discovered another rare survivor of predation. The near-victim was a butterflyfish (*Heniochus*). On its chocolate-brown body were several sets of teeth marks, as if a predator had repeatedly adjusted its grip. Perhaps the hunter was interrupted, or the butterflyfish managed to break away. Terrestrial predators, even dogs or cats, will loosen their bite-grip momentarily to achieve a better hold. A desperate movement during this very brief respite might be the only hope for the victim's survival.

Two sites north of the Calamian Islands (Busuanga) invite special attention because of their pelagic populations. These two, Merope Reef and Hunter Reef, stand alone in open water. Hunter is merely a shallow which never breaks the surface. Like many such open-water outposts, these sudden shallows attract concentrations of pelagic wanderers and are sites for exciting diving. At both Hunter and Merope large sharks, schools of amberjacks and blue jacks, barracuda, tuna, and other blue-water species soar past the diver.

One of the prime underwater wilderness sites in the central Philippines is Apo Reef. Apo has a small island (with a lonely lighthouse), the only land in a much larger reef complex that rises from deep water much like Hunter and Merope. Sculptured stony corals, gorgonians and soft corals, large dome corals, and hundreds of allied invertebrate species crowd together in Apo's shallows and coat its plummeting drop-offs. Such a rich coral population inevitably helps support a large number of reef fish. The corals and mirror-bright reef fish shine against the sober background of gray sharks, silver jacks, tuna, green turtles, and other pelagic passersby. Apo is controlled by the Philippine Navy, which stations two wardens there to protect the reef from fishing, particularly dynamite fishing.

Further north, at the northwest corner of the large

island of Mindoro, lies Lubang Island. When World War II ended, seven Japanese soldiers were inadvertently left behind, well-armed and equipped with survival gear. As the years went by, these fugitives lived by stealing food from local farmers and fishermen. In 30 years this band reportedly killed nearly 350 natives of this rather small island. When one of the last two Japanese was killed by enraged locals, the Philippine government urged Japan to send the remaining soldier's family and military superiors to press for his surrender. Needless to say, he had to be protected by armed escort when he was returned for rehabilitation in Japan.

The diving off Lubang is similar to the diving at neighboring reefs to the south. The same species abound, though the shallow coral gardens are not as well developed as at Apo or the Calamians. Offshore are rather gradual coral-filled slopes going to a depth of 15 meters. At this depth begins a steeper slope, whose base lies at 50 meters. Thereafter the coral thins out markedly, and another gentle open-sand slope continues on toward the open sea.

A rich marine spectrum apparently is present everywhere in the Philippines. In many ways these 7,000 islands represent a tremendous underwater wilderness that is only now being discovered.

215. *A colorful soft coral jutting a full meter upwards from the reef.* (*Gadji Pass, Isle of Pines, New Caledonia*)

216. *A large flounder,* Bothus mancus, *poised for flight from a coral head on the reef slope off Rangiroa.* (*French Polynesia*)

217. *The angelfish* Genicanthus watanabei. *It has found evolutionary advantages in physically and behaviorally resembling a damselfish.* (*Bora Bora, French Polynesia*)

218. Arothron nigropunctatus, *the yellow pufferfish.* (*Rangiroa, French Polynesia*)

219. *The spotted sea bass* (Epinephelus tauvina). (*Rangiroa, French Polynesia*)

220. Epinephelus fasciatus, *the scarlet sea bass.* (*Rangiroa, French Polynesia*)

221. *The hawkfish* (Paracirrhites hemistictus), *nestling among protective coral processes.* (*Rangiroa, French Polynesia*)

222. *A pair of small pufferfish,* Canthigaster valentini. (*Tahaa, French Polynesia*)

223. Centropyge loriculus, *the flame angelfish.* (*Tahaa, French Polynesia*)

224. *The lemonpeel angelfish* (Centropyge flavissimus) *with two bristletooth surgeonfish* (Ctenochaetus striatus). (*Raiatea, Society Islands*)

225. *A vividly colored hogfish* (Bodianus bilunulatus). (*Bora Bora, French Polynesia*)

226. *The bird wrasse* (Gomphosus tricolor), *named for its bird-like motions as well as for its elongated snout.* (*Tahaa, French Polynesia*)

227. *The squirrelfish* (Flammeo sammara). (*Rangiroa, French Polynesia*)

228. *The blue-spotted grouper* (Cephalopholis argus). (*Rangiroa, French Polynesia*)

229. Tridacna *clam. It occurs in great numbers in the shallow waters around the Polynesian islands.*

217

218

219

220

221

222

223

224

227

225

226

228

230. A diver soaring above a rich coral reef off Rangi-roa Atoll. (*French Polynesia*)

231. *A sabellid worm growing on the wreck of the* Heian Maru. (*Truk Lagoon, Micronesia*)

232. *Colorful marine organisms covering the wrecks of Truk in rich profusion. Here a bryozoan, sponge, and soft coral crowd together.* (*Micronesia*)

233. *Anemones perch upon the skeleton of a wire coral growing from the hull of the* Shinkoku Maru. (*Truk Lagoon, Micronesia*)

234. *The mushroom coral* Fungia. (*Palau, Micronesia*)

235. *This exquisite soft coral* Dendronephthya *projects from the superstructure of the* Shinkoku Maru. (*Truk Lagoon, Micronesia*)

236. *A diver examining exquisite soft corals* Dendronephthya, *growing on the hull of the* Rio de Janeiro Maru. (*Truk Lagoon, Micronesia*)

237. *A school of striped catfish* (Plotosus lineatus). (*Lagon d'Erakor, New Hebrides*)

238. *A large gorgonian.* (*Gadji Pass, Isle of Pines, New Caledonia*)

239. *The batfish* (Platax teira) *hovering in the open water above the wreck of the* S.S. President Coolidge. (*New Hebrides*)

240. *An intricate nudibranch moving along a cannon barrel on the* S.S. President Coolidge. (*New Hebrides*)

241. *A nudibranch,* Coryphella rubrolineata, *browsing on algae on the hull of the* S.S. President Coolidge. (*New Hebrides*)

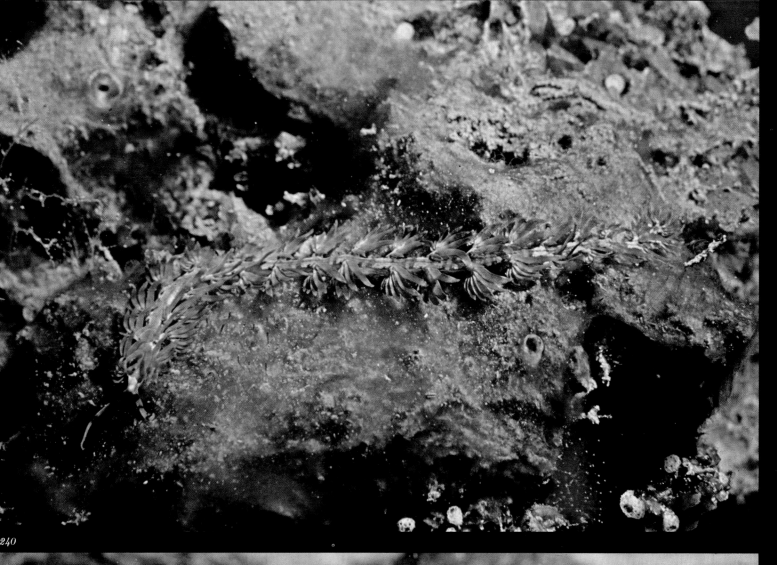

242. *One of the richly hued crinoids of the Philippines. Also known as feather stars, they are believed to be an import from tropical Australia.*
243. *The shallow-scalloped coral* Sarcophyton. *Small animals, such as egg cowries, often use these corals as a perch.* Sarcophyton *corals are widely distributed throughout the South Pacific.*
244. *The soft coral* Dendronephthya *in the midst of lush growth on a Philippine reef. A rich marine spectrum apparently is present everywhere in the Philippines.*
245–247. *A selection of beautiful crinoids of the Philippines.*

The Southern Continent: Australia and the Coral Sea

Australia is a vast island continent southeast of Asia, between the Pacific and Indian oceans. It ranges over 4,000 kilometers from east to west and 3,600 kilometers from north to south. The country is a low, irregular plateau with enormous deserts in the north and west. Southeastern Australia is by far the most fertile and most populated area of the continent. For many centuries, while most of the world progressed through greater and lesser civilizations, Australia was an unknown land populated by a few Stone Age tribes. The natives were dark and wiry nomads, living off a land they shared with such unique inhabitants as towering blue-gum eucalyptus, kangaroos and wallabies, koalas, dingoes, platypus, emus, lyrebirds, and kookaburras. The early Australian tribes are thought to have lived in isolation for at least 12,000 years before the continent was discovered by Europeans.

In the 17th century, Europeans were thrusting toward Indonesia in search of spices to preserve and flavor their meats. A Portuguese captain, blown off course, discovered the new land in 1606. The seafaring Dutch explored it soon thereafter and established a short-lived colony in western Australia known as "New Holland."

It remained for the British to claim and settle the enormous new territory. In 1770, Captain James Cook pioneered Britain's empire in Australia. Sailing along the coast for 2,000 kilometers northward from the inlet he called Botany Bay (after the profuse growth nearby), he claimed all the lands he encountered for Great Britain. A large proportion of the early British settlers in Australia were convicts, and the hostile land was conquered only grudgingly in a history similar in many ways to that of the American West.

In an event that much later took on significance for the underwater wilderness, the British captain Mathew Flinders was dispatched to explore the perimeter of the continent in 1802–1803. In the southern waters, off the coast of today's state of South Australia, Flinders reported seeing huge numbers of great sharks. These legendary sharks have ever since been associated with the cold and stormy waters of the Great Australian Bight. Here, where enormous cliffs rise 300 meters above a wave-lashed coast, the visitor looks out over the sea Flinders described. Cold, green, and forbidding, it is nevertheless rich in life: rays, turtles, squid, and pelagic fish move in cold darkness, the perfect arena for the great shark of the British mariner's account.

In modern times the type of huge shark that Flinders saw has been identified as the great white shark (*Carcharodon carcharias*). Few animals on earth have achieved a more sinister reputation in the popular press, or have so caught the public imagination (Illustration 279). Only a handful of people are known to have been in the water with this shark and returned to describe it. Over the years, bathers in

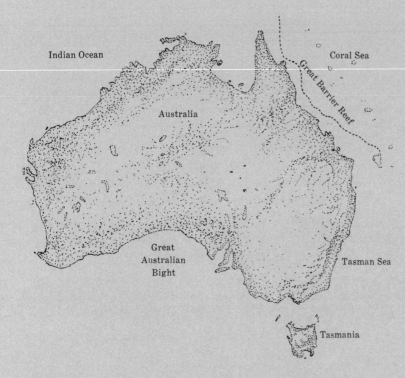

Australia encompasses tropical reefs off its northern coasts and the waters of the "roaring forties" (as the latitude is known) of Antarctica to the south.

Australia, South Africa, and the American Atlantic and Pacific coasts have been attacked and killed by white sharks. In South Australia, three leading spearfishermen were attacked in three separate incidents —and survived. In two cases the victims were in spearfishing competitions and were attacked in water filled with other divers, as well as speared and bleeding fish.

White sharks prey by stealth on sea lions, seals, and porpoises in waters that are usually not clear. Like many sharks, great whites will also attack almost any wounded fish and will eat food wastes from ships. Full understanding of shark attacks on humans is difficult to attain because of the difficulty of studying sharks in their natural environment. There is reason to believe that at least in some cases human intruders inadvertently resemble the sharks' normal prey. Even more important, however, these sharks may be exhibiting behavior characteristic of many species in the wild. Since these large animals consume sizeable quantities of food within their hunting range, it would not be surprising to discover that they warn off possible competitors with a punishing but informative bite. Both of these conjectures are consistent with the pattern of so many white shark attacks on humans. There is an initial mauling after which the shark withdraws.

All of the legends of the great white shark's stealth are confirmed by my experience. In one expedition we encountered nine of these animals in some of the most dramatic diving I have ever witnessed. Since the sharks were our photographic subject, the expedition was equipped with a massive amount of bait and specially constructed shark cages. Baiting for white sharks is an involved, tedious, and tense business. The first ingredient, known as chum, is a thick broth of scrap tuna from the fishing fleet and whale oil flensed from beached, dead whales. The chum is poured into leaky buckets, from which the oily attractant slowly seeps into the water. The resulting slick can be carried by tides for 10 kilometers or more. Almost any white shark crossing the slick will turn upcurrent and home in on the source. Sometimes they are visible for several hundred meters as their dorsal and caudal fins cut the oily, calm surface of the slick. When the shark reaches the boat, it is curious but not aggressive. Now the baiting begins in earnest. Tenkilogram chunks of meat are tied on ropes and hung from floats near the boat. The tension builds as the shark, ever cautious, passes near the bait. After an exploratory pass or two, the shark takes the bait. Lashing its tail and sawing the rope with its razorsharp, triangular teeth, it churns the water. Suddenly, with a flick of the tail, it is gone, and we know the rope has been severed. Again baits go into the water, and again the shark makes cautious passes culminating in a sudden attack; another ten kilograms of meat is invested as an incentive for the shark to return.

After several such feedings over a period of perhaps an hour, the shark cages are slipped into the water and the divers climb into the open top-hatches. There are few sensations to match the anxious anticipation of that moment. The water is cold and green, with a visibility of 10 meters or so. The cage is painted a bright orange, a color that experiments have shown will draw the aggressive attention of sharks. Considering that the shark swimming around us is five meters long and weighs nearly 1,000 kilograms, any shark cage would seem flimsy.

Without warning the shark is there. It is awesome: a torpedo-sharp nose, a great round body, a lashing of its huge caudal fin, and it has come and gone. We resume watching on all sides, swiveling this way and that, hoping to film its entire approach. Bump! It has come up under our bobbing cage and the cage has landed on the creature's back. The muscular tail seems as broad as an elephant's back as it lashes, and the cage is buffeted like a cork in the backwash of the shark's abrupt exit. After such a brush with the cage, the shark may leave for 15 minutes to an hour or more.

Inexorably, the chum draws the great animal back, and once again it appears with stupefying suddenness. The shark seems to come just when you think that it has gone for good. In the course of three or four hours of working, the shark may make dozens, even hundreds, of long, graceful, unhurried passes. Since several chunks of bait are set simultaneously, the shark may approach from any direction. I have seen sharks make a vertical approach, directly upward from the depths, take the bait, and stand on their lashing tail as they fought the rope.

Some of the baits, hung from small floats in front of the cages, were attached to ropes which led back to the boat. When a shark would take one of these baits, the bait boat crew (we had two boats) would haul on the rope, pulling the thrashing shark toward our cages for close-up pictures. A few times the sharks passed within a meter of the cages at high speed. One of the most memorable features of these white sharks is their eyes. They are large, flat, and black, and they seem always to be focused on you, the observer. Sometimes as the shark passes a meter or two away, it enhances its impressive stare by rapidly opening and closing its maw. One's imagination readily supplies the sound effects of gnashing teeth. When the shark commits itself to bite, it frequently rolls its black-disc eyes completely back out of sight. Since the shark eats prey which sometimes thrashes wildly, this would appear to be a wise adaptation to protect the all-important eyes from damage. Once we observed a large white shark attack a sack of tuna chum, ripping and thrashing so blindly that it cracked its head against a camera held by one of our photographers.

This eye retraction gives swift prey an advantage. Several times we pulled bait from the path of a shark committed to bite, and the shark proceeded blindly to bite the spot where the bait had been. This has led us to speculate that if a diver could spot the shark during its approach, he might be able to take some evasive action while its eyes were retracted.

One day, leaving the bait boat to chum for sharks, a few of us cruised some 10 kilometers away to a small island. Here a school of blond-haired sea lions cavorted in shallow water near a beautiful sandy beach. We entered the water with mixed feelings, for we knew these creatures were perhaps the favorite prey of the white shark. Diving and filming in water only four or five meters deep, we were soon surrounded by female sea lions who nuzzled us and pirouetted playfully. Throughout our dive, however, we were haunted by an account given by one of the Australian attack victims. He, too, had been enjoying the antics of a large school of sea lions when he suddenly found himself alone. Before he had grasped the implications, a large white shark took him from behind, severing his leg. For the first time in many dives with sea lions, I was hard-pressed on this sunny day in South Australia to relax and enjoy their play.

When we returned to the boat the radio was crackling —white sharks had appeared at our bait boat. Racing back, we found the bait boat crew well along in feeding a shark. Soon a second shark joined the first. The photographers jostled each other on oil-slick decks to film the feeding sharks. Then the cages were put into the water and the action accelerated. A third shark appeared, and for five hours one or another of these majestic animals swept repeatedly past, devouring bait, nudging the cages, biting the boat propeller and rudder, bumping the hull with their noses and generally putting on an awesome performance. In several of our photographs from this episode the shark's nose is stained blue with paint from the bait boat's hull (Illustration 279).

I feel that these magnificent animals are the victims of hysterical reportage. Far from being the savage, vengeful menace portrayed in the news media, they are merely supreme, lordly predators hunting in their own realm. When people venture into murky waters in seal-like suits, and particularly when they spear fish or hunt abalone, they are unwittingly creating both a feeding attraction and a territorial intrusion for this species. Since the shark is adapted to seeing when the visibility is poor, its lateral-line sensors may bring the hungry predator within striking distance before it even sees its victim. In that final instant, subtle distinctions between a rubber suit and sealskin, or between a speared fish and a fisherman, may become blurred.

It is also possible that a diver or even a boat may well draw territorial aggression by merely entering the water within the range of a large white shark. Shark attacks on boat propellers or shark cages show the clear intention to bite. Indeed, I have observed great white sharks biting a boat propeller when fresh bait was hanging in the water two meters away. I have also seen one pass through a forest of dangling bait to ram the boat just beneath where I was standing. It seemed clear in these cases that hunger played no part in the shark's action; he was punishing an intruder.

This species of shark has presumably remained unchanged for countless millions of years. It has always hunted, and has defended its range in the same way. Our forays into the sea sometimes fail to take that history into account.

My own encounters with this enormous predator left me with a sense of awe and respect. For its power, for its uncanny stealth, for its very perfection at what it has evolved to do, this is one of the great creatures of the underwater wilderness.

Tropical Australia: The Great Barrier Reef Province

So vast is the Australian continent that while white sharks roam the cold "roaring forties" (as its latitude is known) of its southern coast, tropical seas wash its northeastern coast. Here in the hot Pacific sun, the view from the shore encompasses one of the underwater wilderness's most famous communities, the Great Barrier Reef of Australia. Sparkling like green and aquamarine jewels in the cobalt blue of the sea, the islands and cays of the great reef are scattered across an area of 250,000 square kilometers. Captain Cook was the first of the British mariners to explore the incredible intricacies of these coral expanses. In fact, Endeavour Reef in the Swains Group was named for an incident in which Cook's ship went aground and was freed only by a subsequent, fortunately higher, tide. Even Cook, that superb master of seamanship, could be fooled by the subtleties of these uncharted reefs. To this day, many sections of the reef are off-limits even to Australian military vessels because of the hazards of the underwater obstacles, and because it is impossible to keep navigational charts up to date as the reefs shift and grow (Illustration 28).

Despite all the lore of the Great Barrier Reef, there are some persistent misconceptions about this most beautiful of tropical habitats. Most important, it is neither a single reef, nor even a series of reefs. It is also not a "barrier reef" as that term was proposed by Charles Darwin. The Great Barrier Reef is in fact an extraordinary complex of high, tree-covered rock islands, coral cays, and sandspits breaking the surface; coral structures beneath the surface but plainly visible; and even coral atolls hundreds of kilometers out in the Coral Sea. These highly varied coral, sand, and rock structures all occupy the waters of the continental shelf of Australia, and all rise above the 100-fathom line.

Scientists now tend to describe this vast area off the Queensland (northeastern) coast of Australia as the Great Barrier Reef Province. The Great Barrier Reef Province stretches from 9 degrees S to 24 degrees S latitude. Its width ranges from 20 or 30 kilo-

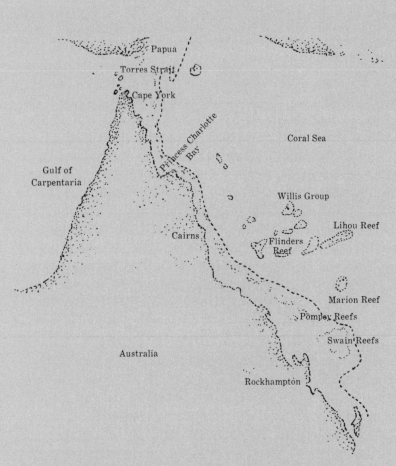

The Great Barrier Reef Province is famous for its quarter million square kilometers of highly variegated islands and multitudinous reefs.

meters to nearly 300 kilometers. Practically the entire area has the conditions of water temperature, sun penetration, and nutrient flow that enable a large variety of coral species to flourish. At the southern end of the province, the waters fall below 18° C and no longer support coral growth. At its northern end the heavy silt and turbidity of New Guinea's Fly River suppress coral development. Between these two extremes, and particularly in the crystal waters of the outer reefs on the Coral Sea Platform, flourish some of the most brilliantly colored and highly variegated marine fauna seen anywhere.

There are some distinct differences between the reefs of the Great Barrier Reef itself and the open-ocean reefs of the associated Coral Sea Platform. Immediately apparent is the water clarity. The Great Barrier Reef proper is rather shallow, and water circulation is impeded by the enormous number of reefs and cays. Thus underwater visibility tends to be on the order of 10 to 30 meters, though it is clearer as you approach the outer ocean edge of the reef. At the atolls of the Coral Sea Platform, such as Marion and Lihou Reefs, the many openings in the ring-reef allow the central lagoons to be flushed frequently by clear ocean water during tidal movements. Here we find the most wonderful confluence of conditions for the undersea observer: the immense riches of the Great Barrier Reef fauna in waters so clear that visibility reaches 70 to 80 meters.

The most obvious of the thronging faunal groups in the Great Barrier Reef Province are the corals. They are here in unbelievable numbers and varieties, and their size is awesome. In the oceanic reef lagoons they have built pinnacles as large as a building of 15 to 20 stories. Dozens of these huge underwater cities of coral are scattered about the lagoons.

It is almost impossible to convey one's impression on first seeing these structures. When I thus saw them, the sunlight was streaming down through the clear water like corridors of light in a lofty cathedral. Swimming away from our cruiser were the small sharks that a few minutes earlier had rocketed to the surface at the clatter of our anchor. Now they cruised the open water with majestic grace, restless scouts on an endless search.

Then, at the limit of visibility, loomed the coral towers that we had come so far to see. Rising 30 to 40 meters from the floor of the enormous lagoon almost to the surface, they were breathtaking. As we drew near we could see that these "bommies" (from the Australian aboriginal word "bombora," a coral structure that does not break the surface) had the same rich community of life that festoons the Great Barrier Reef structures 200 kilometers to the southwest. Baroque spires, domes, and plates of living coral colonies on the towers seemed to tumble over each other in a frantic reach for the sun. Clusters of crinoids spread their delicate arms; in their intense yellows, reds, and greens they outshone even the

crinoids of the Philippines. All around were clouds of fish of every description: tiny damselfish, angelfish, and butterflyfish hovered near the protective crevices of the sunken metropolis, while in the blue water some distance away the foxes and wolves of the sea, amberjacks and barracuda, patrolled patiently. Here and there in the coral towers, gigantic cracks and crevices provided contrasting zones of darkness and shelter to the gaudy show of life on the pinnacles. In these darker retreats, clouds of small anchovies and cardinal fish hovered, large fans of canary-yellow gorgonians branched, and delicate, water-inflated soft coral colonies shone in copper and vermilion. Small animals such as the egg cowries (Illustrations 261–263) perched atop (*Sarcophyton*) shallow-scalloped corals (Illustration 243), gently munching the plentiful polyps. Long, wormlike nudibranchs (shellless snails) reached up like miniature giraffes into the branching coral to feed. In one valley we saw a huge red nudibranch decorated with frilled cocoa minarets. This conspicuously decorated animal, whose gills (which are the frills we see) are external, blended into the lacy branches of the gorgonian upon which it was feeding. When carried up into the open water, the nudibranch's gills acted as sails to carry it to a safe landing back on the reef-top.

Elsewhere on the coral parapets, gaudy sea cucumbers were draped over small coral projections, above the sheer drop to the lagoon floor nearly 50 meters below. Throughout the maze of life we constantly came upon the iridescent mantles of the giant clam Tridacna, siphoning water continually through their bodies.

At night in the sandy shallows, just within the outer reef, great armies of shellfish move about, leaving their characteristic trails in the sand. Augers, mitres, volutes, cowries, and other delicate mollusks pick morsels from the sand. One can even find cowrie shells which have been appropriated by hermit crabs. Often atop such shells are small anemones, their tentacles waving like the plumes of miniature knights. On shallow reefs one may also discover the asses ear shell (*Haliotus asinina*) browsing, its tender lettuce-green body far too large to fit beneath its shell. The butterflyfish of the Great Barrier Reef Province are as colorful as any in the world; triangle, banner, right-angle, masked, beaked, long-snout, and dozens of other varieties of this fish pursue their rounds, picking a bit here, a bit there. The great reefs serve not only as a feeding ground for these brightly-garbed fish, but also as a refuge. One of the fascinating aspects of this sanctuary is that the retreat of the little fishes is exactly calibrated to approaching danger. If one gently glides in to take a picture, they glide just as gently away, always a short distance out of comfortable camera range. Similarly, when one tries to lunge in to photograph them, they bolt away. Some species, such as the cardinal fish (*Apogon*), are extremely wary, eternally poised for flight.

Other Australian reef species protect themselves in different ways. One may spend hours watching the clownfishes (*Amphiprion perideraion* and *A. unimaculatus*) dancing lightly among the tentacles of their host *Stoichactus* anemone. Often more than one species of clownfish will live in a single large anemone, diving in and out among the armed tentacles. At the approach of human intruders the clownfish will usually disappear into the writhing mass of tentacles. Since some anemones tend to retract their tentacles completely when feeding, one may observe the unusual situation in which the clownfish are literally dispossessed, bereft of the shelter the tentacles usually provide.

The balance on any reef is signified by the health of all its participants. In an isolated reef such as those on the Coral Sea Platform, there are such multitudes of healthy specimens from so many species that the proper functioning of the entire system seems obvious. There are no massive areas of dead and dying coral, no fish with diseased scales and frayed fins. There is not even the much-publicized crown-of-thorns starfish. On the healthy reefs of the Coral Sea Platform, only an occasional crown-of-thorns can be found. The few that are present are large and scattered about the pinnacles. Since the crown-of-thorns digests coral polyps by extruding its stomach and flooding the living coral with powerful digestive juices, it is occasionally surrounded by the bleached skeletons of consumed coral. Each of the major coral pinnacles maẙ have a few big starfish among its denizens, but the overall coral growth is simply too vast for those few to affect the total scene.

On the Great Barrier Reef proper, the situation is rather different. Several thousand square kilometers of coral have been badly damaged by a population explosion of crown-of-thorns starfish. There is much speculation and intense research into the infestation, which may be no more than a periodic outburst that has already begun to reverse itself.

Not only the crown-of-thorns but their principal predator, the triton trumpet (*Charonia tritonis*), are found in relative abundance on the outer reefs of the Coral Sea. These large mollusks have been heavily collected as souvenirs in the more accessible areas of the Great Barrier Reef. Some observers believe that the collecting has reduced predator pressure on the population of the crown-of-thorns and contributed to the outbreak.

Among the most fascinating inhabitants of the tropical Australian reefs are the pelagic species and larger predators. Gorgeously patterned moray eels such as *Gymnothorax meleagris* spend their days resting in coral crevices and their nights hunting. Large and colorful groupers hover in the shadows of coral heads, studying their human visitors curiously. Often while photographing small invertebrates among the coral, I would become aware of a grouper staring over my shoulder. As I turned ever so slowly to look at it, the first eye-contact invariably resulted in its sudden departure.

Most of the original Englishmen who settled Australia were convicts and their jailers, unsophisticated men who were familiar only with the waters of their distant homeland. The only fish these men had ever known were cold-water species such as cod or the trout of the English lakes; there are no trout or cod on Australian coral reefs. When the Britons came to this far-off land and saw the native fish, it was perhaps natural for them to bestow on them names from home. Thus, even today, common groupers in Australian reference works are called "coral trout" or "coral cod" (Illustration 272). The most outlandish of these misnomers is surely that given the gaudy lionfish *Brachirus zebra;* this ornate species is known in Australia as the butterfly cod.

In the open waters between the reefs, schools of jacks and tuna pass silently, shadowed frequently by sharks. Large turtles are also abundant in many areas, where they come ashore at night to lay their eggs. One day in the Swains Group of the Great Barrier Reef I was confronted by a turtle well over a meter long. Swimming directly up to me, it hesitated, drifted closer, and then in a great rush turned swiftly and swam some 20 meters away. There it paused and circled back. Once again we came within a meter of each other, and again there came that seeming shock of recognition, flight, and hesitant return. After four repetitions of this strange behavior, the turtle changed its approach. In one of my rarest moments in the sea this great reptile swam lazy circles about me a mere meter away (Illustration 270). Swimming with it, sometimes under its chin and sometimes beneath its plated underside, was a pale yellow remora. Each time I took a photograph I would wait several seconds for the strobe light to recycle. During these intervals I reached out and rubbed the turtle's head to remove an algae growth there. After what seemed an endless communion between us, the turtle swam off into the distance with lazily relaxed strokes.

A year later on a reef in the Coral Sea, we came upon another interesting local inhabitant, a good-sized pufferfish. We first found this roly-poly fish hovering a few centimeters off the sand in a small valley. Laying down my cameras I came slowly near the puffer, which made no attempt to escape. As gently as I could, I took him in my hands and pumped his body. Almost all pufferfish respond to such treatment by inflating their bodies with water. Soon we had a sizeable living balloon on our hands, which we proceeded to photograph. After a while we released the animal, which now expelled its inflating water and began swimming away. Moments later, having completely resumed its normal profile, the puffer disappeared from view.

Before traveling to Australia I was warned of two reef inhabitants generally considered to be highly dangerous: the sea snake, paddle-tailed relative of the terrestrial cobra, whose venom is highly toxic, and the shark. One of the main impressions I have garnered from several trips to Australia is that a healthy reef environment satisfies the needs of its inhabitants, and so its predators pose little threat to human visitors. While the sharks and particularly the sea snakes exhibited great curiosity, often approaching close enough for easy photography, I rarely felt threatened. It is true, however, that the presence of large numbers of these predators makes one's first dives in Australian waters highly charged events.

The sea snakes I observed were completely peaceful, spending most of each day lying about in sleep (Illustration 276) and hunting at night for small fish sleeping in coral crevices. As breathers of air, the sea snakes must periodically go to the surface. The time between these excursions for air varies from a half hour to two hours, depending on the snakes' interim activity. Something deep in every human being seems to churn with fear at the first glimpse of a two-meter-long snake boring upward through the clear water. Since sea snakes exhibit great curiosity, the diver's anxiety is often compounded when a snake leaves its flight path and heads straight for him. The fact that snakes have rather poor eyesight can make for some nerve-wracking moments. A snake can also become very aggressive if, for example, a large anchor drops next to it as it sleeps. On one occasion a one-and-a-half-meter olive snake (*Aipysurus laevis*) completely encircled the waist of my diving partner. Another diver suddenly discovered he had a brightly banded sea snake around his neck, staring curiously into his face plate. Snakes have examined my camera, hands, feet, and face mask so thoroughly that it took some self-control not to beat a hasty retreat. That could be dangerous, since many animals will attack at any sign of flight.

The toxin of the sea snake is up to ten times as potent as a cobra's. It is injected through tiny fangs well back in its upper jaw, and is pumped from venom glands. A rubber wet suit used by divers will usually prevent the fangs from piercing the skin. Many of these snakes have developed an extremely efficient hunting method; they swiftly bite, envenomate, and withdraw, then wait for the venom to kill the prey. When the prey is dead, the snake swallows it head-first without resistance. Since snakes eat such dangerously poison-spined morsels as scorpionfish, such a feeding method has proven most efficient.

The sea snake's reputation as a lethal killer must be balanced by a proper evaluation of its temperament as well as of its armament. It seems clear that the injection of venom is a feeding rather than a defense mechanism. I know two men who were bitten while handling the beautiful *Astrotia stokesii*, an extremely potent sea snake, and who suffered only intense anxiety and the discomfort of a sharp-toothed nip. It is very likely that this bite was a "blank bite," essentially a warning nip without envenomation. By nipping rather than killing in routine territorial encounters, the animal's reputation is spread throughout its range. If the snake killed in every encounter, there would be no survivors to transmit knowledge of the snake's aggressiveness or potency. In all my encounters with sea snakes, I have discovered them to display what seems to be great curiosity. Unless handled roughly, these snakes pose no danger to divers. Indeed, encounters between humans and sea snakes are almost invariably benign and often quite beautiful.

My encounters with the plentiful shark population have been similarly tranquil. In general, the sharks (Illustration 278) one encounters in the Great Barrier Reef Province are small reef white-tips (*Triaenodon apicalis*) and small gray whalers (*Carcharhinus amblyrhynchos*). They are seen singly or in small numbers on practically every dive. On numerous occasions the small gray whalers rocketed to the surface on hearing the noisy racket of our anchor chain, then accompanied us at a discreet distance throughout our dive. One day while ten of us were snorkeling in the shallow water above a pinnacle in Marion Reef lagoon, a tiger shark about four meters long soared right through the middle of our group and continued on its way. Another time a five-meter whale shark swept right under our anchored boat as we watched from the deck. In all our contacts with Coral Sea sharks, the mood has been peaceful, because nothing had occurred to provoke the sharks. Low-frequency vibrations in the range emitted by the thrashing of a wounded fish are a powerful stimulus to sharks. For example, I have waved my hand violently from a cage and drawn a white shark directly toward me. Military sources have found that the beat of helicopter blades on the surface of the water have drawn shark attacks on downed pilots.

With this in mind, I have conducted filming expeditions on two occasions to remote Coral Sea reefs where there are known concentrations of gray whaler sharks. On the first trip, we began the day by diving without incident among the ceaselessly patrolling sharks. Without a feeding stimulus, the sharks merely moved from one sector of the reef to another ahead of us. Back on our boat, we decided to incite the sharks so that we could film them during feeding. We went into the water with our Australian guides, baiting the sharks by spearing fish. Within 15 seconds of spearing the very first fish, the shark pack increased in numbers and began swimming more rapidly, casting about for the source of the vibrations. Then one shark found the speared fish, took it in its mouth, stood vertically on its tail, and shook the fish to pieces. There was a stunned pause; then the pack went wild, whizzing about us like cannon shells (Illustration 277). They became so agitated in the lust for food they literally caromed off each other and whipped between us with only centimeters to spare. Twice I fired my strobe light at streaking sharks, not to photograph them but to turn them away before they hit me.

After a few fevered minutes it was over, and the sharks resumed their ceaseless patrol. Our impression after the dive was that the sharks had never hit us because in some way they distinguished between us and the bait; and, even more, that they perceived us as possibly dangerous and avoided us.

I returned two years later to re-enact that scene. In the morning we once again dived with the gray sentinels, filming without incident. After our divers had returned to the boat, I found that I still had some air left in my scuba tank. The water was clear and sun-splashed, so I asked our captain to throw some fish carcasses from the morning's catch over the side. He did, and I hovered below the stern watching the sharks become increasingly agitated in their rushes at the slowly falling scraps. Soon the water was filled with aggressive sharks. I ran out of air and boarded the boat. Below us, the roiling sharks were attacking anything that moved, including an empty yellow film box that had blown overboard.

After lunch, in the early afternoon, I slipped into the water with two large camera rigs. I quickly realized that the current had rapidly increased and I was being carried irretrievably away from the boat. I looked down and saw a chilling sight—25 or 30 sharks milling about below me. As I watched in horror, the mass of sharks formed an inverted funnel and began boring vertically upward toward me. Shouting to the boat to hold the other divers I began back-pedaling across the surface toward the haven of the reef. The sharks maintained formation and streamed after me. In moments I was using both camera rigs to club the noses of sharks snapping at my legs. For a time interval I shall never be able to calculate, I hammered sharks and retreated. Suddenly the attack broke off; I had reached the reef, where, with my back against a solid, welcome coral cliff, I watched the shark pack disperse.

Some images from Australia will remain with me always—the great turtle lazily circling me, the sinuous sea snakes, the rainbow-hued crinoids, the frenzied rush of the Coral Sea sharks contrasted with the deliberate slow grandeur of the great white sharks. But the most serene and enduring memory is of the great coral cathedrals of the outer reefs—their vast stillness, the sunlight streaming down around them, and the perfect sense of completeness that surrounds them. In their pristine isolation, they are one of the purest manifestations of what we seek in the underwater wilderness.

248–278. *A selection of photographs from the Great Barrier Reef, Australia.*

248. *An exquisitely sculpted limestone reef complex, photographed at Gannett Cay in the Swains Group. These coral skeletons are intricate in their close detail, yet spread over meters of the reef.*

249. *A diver photographing a large colony of foliose coral* (Turbinaria frondens).

250. *A ten-armed crinoid taking up its perch on the arms of a golden gorgonian colony. The crinoid competes for food with the polyps of the gorgonian.*

251. *A crinoid on its gorgonian perch.*

252. *Coral Sea crinoids in a rainbow of colors.*

253. *A delicate soft coral* Dendronephthya. *The soft corals inflate their skeletal bodies with water in preparation for feeding. When not feeding, the polyps retract into the surrounding fleshy tissue and the skeleton collapses in a soft lump.*

254. *The clingfish* (Lepadichthys lineatus). *These clingfish are a polychromic species that live out their lives among the arms of crinoids. The body coloration of the fish is specifically adapted to that of their crinoid hosts.*

255. *One of the incredibly varied soft corals* Dendronephthya.

256. *Coral colonies from the Great Barrier Reef.*

257. *A close-up of the gorgonian* Mopsia ellisi.

258. *An exquisite crinoid.*

259. *Polychaete worms* Spirobranchus giganteus. *They occur in particularly brilliant hues on tropical Australian reefs.*

250

252

251

253

254

255

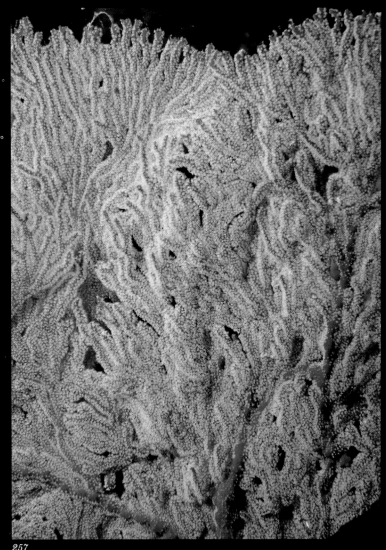

257

256

258

260. *A well-developed reef area featuring an umbrella coral that is sharing its perch with a crinoid and a gorgonian.*

261–263. *Three photographs of the egg cowrie* (Ovula ovum) : *its beautiful shell and the animal in its natural habitat on the reef.*

264. *The leathery-warty nudibranch* Phyllidia.

265. *A nudibranch* Notodoris metastigina *dwarfing the pinpoint-sized coral polyps about it.*

266. *A cuttlefish remarkable for its extraordinary range of color and body texture variation. Of more than 30 photographs taken of this individual, no two are alike.*

267. *The damselfish* Paraglyphidodon behni.

268. *The large angelfish* Pomacanthus semicirculatus.

269. *The beaked butterflyfish* (Chelmon rostratus).

270. *A huge sea turtle. It swam in circles around the photographer, and let him remove some algae growing on its head.*

271. Gymnothorax meleagris, *a spotted moray eel.*

272. *A large coral trout* (Plectropoma maculatum) *bedded down for the night. It rests on its extended fins. On its lip a small cleaner shrimp* Periclemenes *removes parasites.*

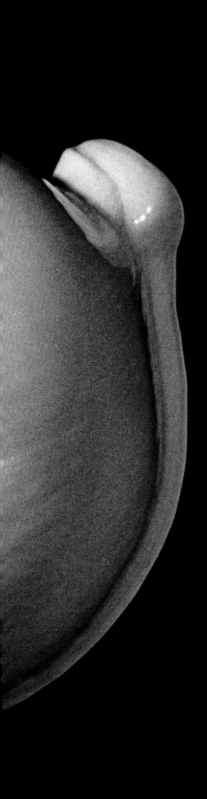

262

263

264

265

266

267

268

269

270

271

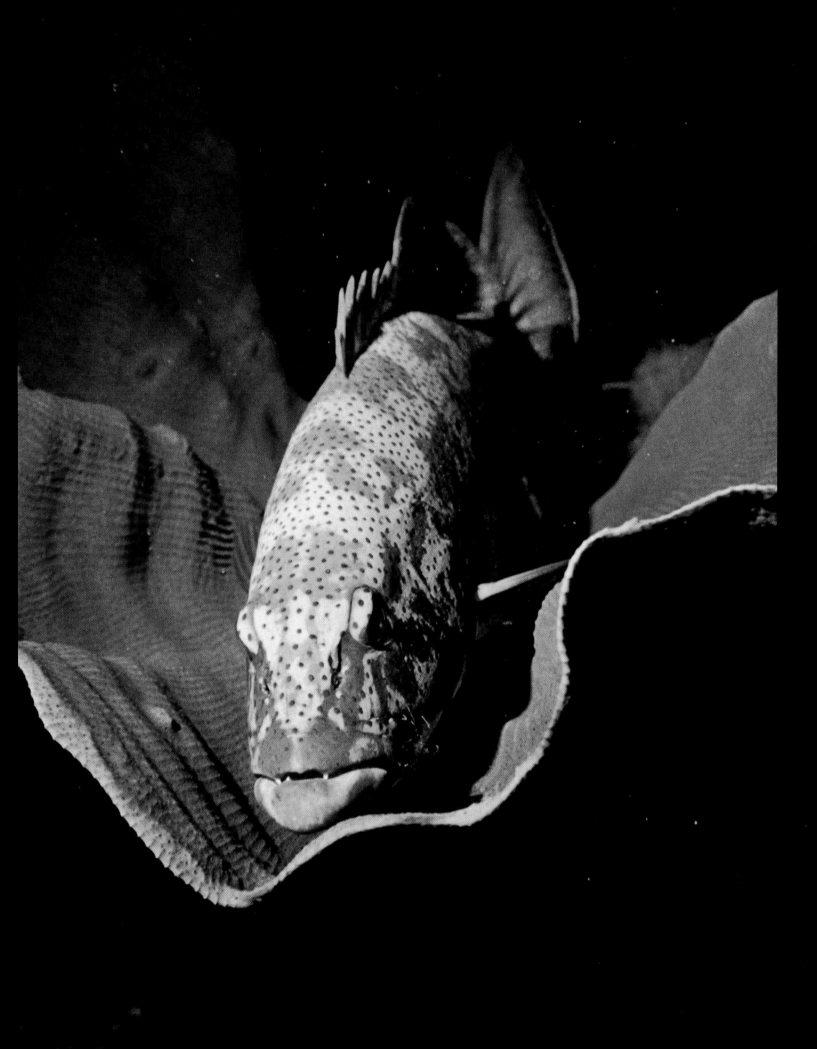

273. *A colony of yellow tunicates.*
274. *The clam* Tridacna maxima *amid* Dendronephthya *soft corals and serpulid worms* Spirobranchus.
275. *The crinoids offer endless color variations.*
276. *A sleeping olive snake (*Aipysurus laevis*).*
277. *The beginning of shark frenzy. These Australian gray reef sharks (*Carcharhinus amblyrhynchos*)
were beginning to make high-speed passes.*
278. *Gray reef shark (*Carcharhinus amblyrhynchos*)
cruising in open water. Few swimming sea animals
are as graceful as these much-maligned creatures.*
279. *The great white shark (*Carcharodon carcharias*), also known as "man-eater," "white pointer,"
and "white death." (Dangerous Reef, South Australia*)

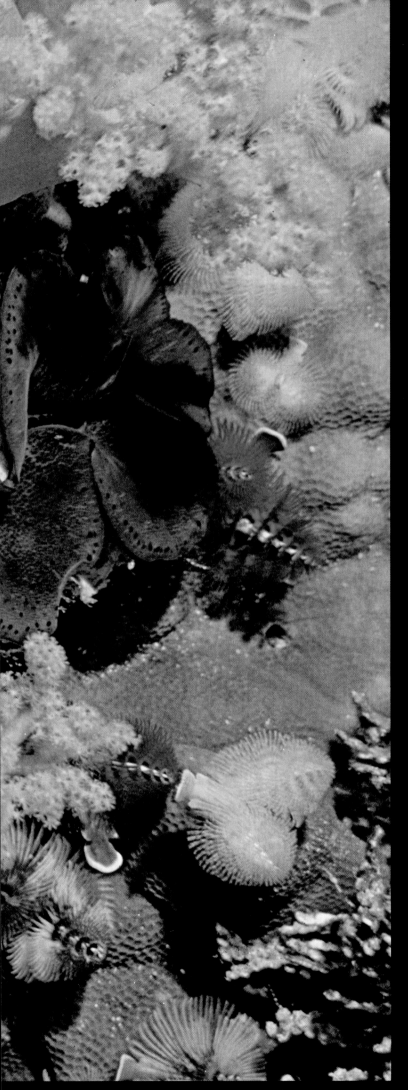

Man's Oldest Seas: The Mediterranean and the Red Sea

In considering these two seas, we search in the oldest shadows of human history. The earliest tribes, the earliest conquerors, and the earliest divers played out their dramas in this geographic arena.

The Mediterranean

At the western end of this nearly landlocked sea lie the straits of Gibraltar, where the mountains of Spain and the deserts of North Africa confront each other across a few kilometers of water. It is here that the Mediterranean and Atlantic are linked, but ever so tenuously; for the geography of the straits is such that the water passage is both narrow and shallow. Thus the waters of the larger Atlantic cannot easily intermix with and refresh those of the enclosed Mediterranean. This geography has serious consequences for a sea so intimately linked with the stormy history of mankind. It means that any harm that befalls either the marine life of the Mediterranean or the sea itself is repaired but slowly by the inflow of Atlantic waters.

With this limiting reality in mind, we might reflect briefly on the millennia of man's activity in this relatively small sea. Many thousands of years ago the races of men gradually increased their numbers, until the scattered hunting tribes of the paleolithic and neolithic men gave way to organized societies. Among the major early cultures that ringed the Mediterranean were those of the Sumerians, Assyrians, Chaldeans, and Egyptians.

The first sea-trading cultures were those of the Minoans and Phoenicians. We know that as far back as 3,000 years ago the Phoenicians explored the Mediterranean, the Red Sea, the Persian Gulf, and the Atlantic coast of Africa. These ancient dominant cultures gave way to the Hellenes—early Greeks—whose own culture flowered in part because of their marine trade. For the Greeks, Romans, and subsequent cultures that centered on the Mediterranean, this sea was a crucial factor in shaping their economic and political circumstances. While peoples fought, struggled, lived, and died along its shores, the sea, though vulnerable, seemed to change very little. But during these centuries, unfortunately, patterns of thoughtless and destructive usage of the sea evolved which unavoidably resulted in dramatic changes in the Mediterranean. In the 20th century, such things as pollution, overfishing, the use of dynamite, and unselective netting have had devastating—and now visible—effects on the sea, especially on its former underwater wilderness regions. The Mediterranean is included in this book, therefore, mainly for its historic importance and as the sea where diving began.

The mountainous coasts of Spain and Italy and the jewel-like islands of Greece are perhaps as beautiful for boating and sight-seeing today as they were a thousand years ago. But underwater there have been great changes. The Mediterranean, at least since the coming of man, has never been as rich in reef

The Mediterranean, the earliest known sea, has been central to the ecological history of mankind.

corals and fish as such tropical areas as the Caribbean or South Pacific. However, it was well populated with resident subtropical and temperate-zone species as well as with a host of pelagic fish. In the 1930s and 1940s, some of the early sport divers hunted the large grouper (*Epinephelus gigas*, called "merou"), the sea bream (*Aurata aurata*, called "gilthead"), wrasses such as the "seawife" (*Labrus bergylta*), mullets (*Mugil cephalus*), corvinas (*Corvina nigra*), and turbot (*Rhombus maximus*).

In those early days, Guy Gilpatric (perhaps the first to dive purely for sport), Philippe Taillez, Hans Hass, Jacques-Yves Cousteau, Albert Falco, Frederic Dumas, and a host of other pioneers plunged into the Mediterranean with makeshift experimental equipment and brought us the first glimpses of the wonders beneath its surface. But even then the deadly pressures on the Mediterranean's sea life were starting to build. The growing army of spear fishermen soon found their newly discovered hunting grounds under savage attack. This development was described vividly by Dr. Gilbert Doukan in his book *Underwater Hunting* (1948):

These places, which are paradisal for the fisher with the harpoon, have their own purgatory: the dynamiters. Already, even on our continental shores, melinite is doing a terrible amount of damage. Here, what with the impunity conferred by isolation, the complicity of the coastal peasantry, and the great distances between the coastguard stations, the devastation is appalling. The professional fishermen no longer trouble to set their nets. At night one hears the dull explosions that tell us they are at work. A shoal of giltheads? Dynamite. A rock surrounded by bream? Dynamite. A hole in which the Jewfish takes refuge? Dynamite. The Corsican coast, like our own, may be cleared of fish for years to come if we cannot put a stop to this destructive folly. In the summer of 1946 we were able to realize the extent of the mischief, for we found veritable deserts where in 1942 the fish abounded in the creeks and inlets.

Since World War II, the damage from two generations of dynamite-fishing, trawling, and the accumulated chemical pollution of an industrial society has rapidly depleted marine life. The lack of flushing action by the Atlantic has resulted in localized pollution levels that seriously threaten the future of this historic sea. In fact, "localized" hardly describes some of the pollution incidents. For example, the freighter *Cavtat* was rammed and sunk in 1974 near Otranto, Italy. Nearly 900 barrels containing tetraethyl and tetramethyl lead are slowly corroding on board the sunken wreck. Sometime in the next 10 to 15 years, this lethal poison will begin spreading death across this nearly enclosed sea.

Today the Mediterranean still has life and color. As our pictures show, delicate gorgonians, hydroids, and nudibranchs share the algae-rich surface of undersea boulders and rock croppings. Here and there the patient diver still finds the *langouste* or lobster. Even corvinas and mullets and wrasses can still be found. It is the comparison of the Mediterranean today to what it was even a few decades ago that awakens our sense of loss. It is not what we find but what we do not find any longer that sounds a warning to civilized man.

In an earlier chapter, we considered the Caribbean's majestic Palancar Reef a sepulcher after two decades of active human predation. The scale of damage in the Mediterranean is immeasurably larger. Where the possible death of Palancar does not in itself pose a threat to human life on its adjacent mainland, the death of the Mediterranean would affect millions. For now, its fabulous sunken antiquities lie in waters whose very future is in doubt.

The Red Sea

At the southern fringe of the same historic arena lies the incredibly rich Red Sea. Essentially an arm of the Indian Ocean, it begins at the Strait of Bab al Mandab, between South Yemen and Ethiopia, and forms a long band running from southeast to northwest for 2,500 kilometers along a great tectonic fault. It washes the sun-baked shores of Ethiopia, the Sudan, and the United Arab Republic (Egypt) on the west, and South Yemen, Yemen, and Saudi Arabia on the east. This main body of the Red Sea marks the northeastern edge of the African continent.

At its northern end, this great fault line intersects another tectonic boundary, the Syrian-African rift, running almost due north to Syria. At this point the Red Sea splits into two gulfs, the Suez (merely a continuation of the Red Sea proper) and the deeper Gulf of Aqaba, also known as the Gulf of Eilat. These two bodies of water border the strategic Sinai and give two more countries, Israel and Jordan, access to the Red Sea.

The Red Sea has periodically served as a critical marine route for commerce between European and Middle Eastern countries and such destinations as India and Africa. Because of its landlocked nature its weather is relatively benign. However, its political climate has not always been calm. Indeed, the history of the Red Sea has been as turbulent as that of the Mediterranean. We know that Eilat was a seaport for King Solomon. The Egyptians, Greeks, and Romans plied the Red Sea regularly, and both the Egyptians and Persians are known to have attempted the building of a canal to link the Nile River and the Gulf of Suez.

One of the great powers of the region 2,000 years ago was a tribe known as the Nabateans, whose great walled city of Petra still stands in brooding hauteur today. At one time these raiders controlled both the sea and the land routes to the east, exacting tribute from all who passed. But the tides of history engulfed the Nabateans. Petra, "the rose-red city half as old as time," was finally invaded by the Romans. Today, the monumental, polished sandstone

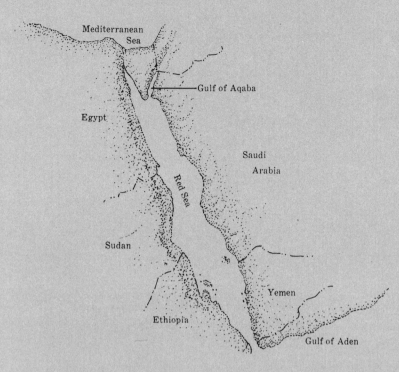

The Red Sea is a lush oasis of marine life surrounded by vast tracts of mountain and desert.

tombs of the Nabatean kings look down on the empty ruins of the Roman temples. The visitor to this spectacular valley finds himself humbled by the grandeur of the place. One can only imagine the glories of Nabatean life 20 centuries ago while watching today's Bedouins encamped in the shadow of the massive tombs.

Byzantines, Moslems, and Christian Crusaders successively controlled the Red Sea, giving way in the 19th century to the Turks of the Ottoman Empire, whose hegemony continued until World War I. British interest in trade with India had been instrumental in bringing about the mighty Suez Canal project in 1869, and the Canal had provided secure trade and military routes for the major European powers. When the Turks joined with the Germans before World War I, the stage was set for confrontation in the desert in 1914–1918.

Onto this stage stepped T. E. Lawrence, known as Lawrence of Arabia, a brilliant young officer in the British Army, intensely devoted to understanding the Arabs. His lightning strikes on camelback spearheaded the English military strategy, and in the end the Turks and Germans were defeated. To Lawrence's distress, victory revived the basic rivalry between the Arab chieftains, and the European powers cynically began carving up the spoils. Modern history has proved Lawrence's dream of a stable Arab leadership to be as elusive today as it was in his time.

Through all of this political and economic ferment, the Red Sea, unlike the Mediterranean, has maintained a viable, healthy undersea society. The principal reasons for this are the relatively sparse population of the countries bordering these waters, and the fact that they are not developed industrial societies. Moreover, the Red Sea is essentially tropical while the Mediterranean is not. Thus it is blessed with that kaleidoscope of color and variety we associate with Caribbean or South Pacific reefs.

The Red Sea has rich concentrations of intricate branching corals, among which the most common and abundant are the umbrella corals and the fire corals. These and several other dome corals form fringing reefs on the sloping stone of the chasm, between tectonic plates, that forms the sea. In some areas the building corals themselves have formed substantial structures: great ridges of coral, perpendicular to the shoreline, and honeycombed with fissures and small crevices that offer shelter to a multitude of fishes and invertebrates. For example, in the intricate branches of the shallow-water fire coral is sheltered the spotted blenny (*Blennius nigriceps*). This shy creature peers out with a huge eye from a white, black-spotted head which crowns a colorful, red-spotted body. Among the same coral branches the observer will find small crabs, shrimp, several varieties of cardinal fish, damselfish, gobies, wrasses, sea urchins, and brittle starfish.

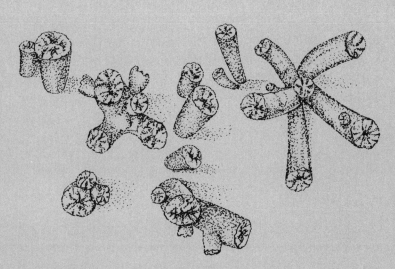

One of the many growth forms of Dendrophyllia *coral.*

The waters of the deeper slope are adorned with occasional stark tree coral (*Dendrophyllia*). These dark green, branching structures with their sparsely scattered polyps may spread a meter or more in breadth, but they offer minimum shelter. A few gobies skitter out of sight around the branches, but otherwise the tree stands quite alone.

Another typical form of Red Sea reef is the open sand-and-eelgrass bottom with scattered small reef structures, each from one to five meters in breadth and height. These small "patch" reefs are commonly surmounted by large umbrella corals, which help camouflage other reef inhabitants in their broad shadows. As might be expected, many species forsake the open bottom and cluster about these patch reefs. Each reef becomes an arena of life, with thousands of varied individuals living in close proximity, bound together by the common need for shelter from predation.

One of the most colorful and numerous of these organisms is the orange fairy basslet (*Anthias squammipinnis*). This lovely and graceful fish is a bright orange with iridescent facial markings, a slashing, long banner on its dorsal fin, and a flowing swallow tail (Illustrations 295–296, 300). Basslets occur in large numbers about most of the larger shallow-water coral heads. An interesting sidelight on the basslets is that each small school will have a dominant male whose color is not orange but a rich purple (Illustration 295). Scientists have found that when this dominant male is removed or dies, one of the females immediately changes both gender and color and takes its place. In the afternoon sun, these clouds of fairy basslets form a shimmering aura about their residential coral communities.

In some areas, gaudy soft corals in dazzling pinks and reds perch atop the hard corals (Illustration 291). The soft corals are very healthy, well nourished in the plankton-rich flow of water over the reef. One species grows in aggregations as much as two meters across. Their bushy colonies offer shelter to a great variety of fish and invertebrates, since they are so thick that one cannot see into their interior. These soft corals, designated *Litophyton*, are characteristic of Red Sea reefs and very abundant in the shallower depths (5 to 20 meters).

The crinoids are another group of intricately shaped residents on these Red Sea reefs. Interestingly, they do not have the incandescent colors of their relatives in Australia, the Philippines, or even the Caribbean. Red Sea crinoids are generally a delicate, unobtrusive black or deep maroon. Some are decorated with patterns of silver or white. Many seem to seek shelter by day in coral crevices, indicating that they may be night-feeders. Other crinoids take up residence among the cloud-soft arms of the soft corals, their own puritan garb a gentle contrast to the corals' brilliant colors. In addition, these crinoids are inhabited by large numbers of commensals, just like their radiant relatives in the South Pacific and elsewhere. Clingfish, galatheid crabs, shrimp, and copepods are found in colors to match those of their hosts in the previously described phenomenon known as polychromism. Some of the smaller commensals are found by the hundreds on each crinoid host.

The larger coral formations of the Red Sea harbor vast numbers of retiring species such as sweepers, anchovies, and cardinal fish. A single large coral head may have thousands of individuals hovering in its shadowy crevices. One might expect to find a milling throng of fish, but each species seems to school by itself. Thus, a human observer entering a cave in a large coral formation and looking outward into the open water through an enormous throng of these silvery shoaling fish may note that each species holds its own formation. One sees a distinctly separate school of sweepers, crossed by a separate school of cardinal fish, and so forth. The members of these schools are so accustomed to their close-order movement that one formation may pass right through another with no physical contact between the members. Little is known of the kind of communication that makes such schooling discipline possible. If this were better understood, it might have significance in predicting the behavior of open-water schooling fishes so important as human food.

The angelfish and butterflyfish of the Red Sea, as elsewhere, are among the most colorful and active of the reef-dwellers (Illustrations 298, 303). One finds a profusion of such old friends as the spectacular emperor angelfish (*Pomacanthus imperator*), the king angelfish (*Pygoplites diacanthus*), and the masked butterflyfish (*Chaetodon lunula*). There are also more localized specimens such as the elusive map angelfish (*Pomacanthus maculosus*) and the delicate-looking red-tailed butterflyfish (*Chaetodon chrysurus*). The latter almost always appear in pairs, daintily picking tidbits from the corals in an unhurried fashion (Illustration 304). The only time I have ever seen these reef-fish become aggressive was when four red-tailed butterflyfish briskly attacked a jellyfish that had drifted down out of the open water and reached the reef. The jellyfish may have already been injured or dead when it came so uncharacteristically close to the reef. In any event, the four butterflyfish converged on it in a small, sandy valley surrounded by corals and soft corals. With short darting motions they attacked, tearing chunk after chunk out of the jellyfish, eating some of the pieces but letting most of them simply drift away. The four butterflyfish then formed two pairs and resumed their rather tranquil swimming and nibbling about the reef.

The Red Sea reefs are alive with hundreds of other fascinating inhabitants. On the open sand one may find the weeverfish, or grubfish (*Parapercis cylindrica*), which wait motionless until you have approached to within a few centimeters, then dart to safety. Another predator that does not move until its prey comes close is the scorpionfish, whose frilled and scaled body resembles the surface of an algae-covered rock. Its mouth is large and peculiarly adapted to the erupting motion with which it engulfs small fish that pass by. This fearsome-looking fish has 12 poisonous dorsal spines, which waders in shallow water and divers have often failed to see until a foot or hand was pierced. The venom, while not as potent as that of the related deadly stonefish *Synanceia verrucosa*, may still cause dizziness, nausea, and severe swelling. Because the scorpionfish's camouflage is so effective, a cautious diver carefully examines the rock upon which he is about to rest his hand.

The ever-present cleaner shrimp, including several species we have encountered in other oceans, is found in small coral crevices all over the reef. The barbershop shrimp (*Stenopus hispidus*), a gaudy specimen with red and white bands and sparkling blue highlights, is widely distributed in tropical seas around the world. *Hippolysmata grabhami*, another shrimp found in many seas around the world, has striped colors of yellow, red, and white running longitudinally down its slender body. Other types of cleaners abound, notably the cleaner wrasse (*Labroides dimidiatus*), which offers its services with a characteristic stylish, bobbing dance.

Coral crevices yield all manner of surprises. One good-sized fish often found hiding by day is the blue-spotted boxfish. This blue and gold fish has a rather homely face and a rectangular, box-like body of fused scales (Illustration 299). Its armored body, while providing extremely effective protection, can, like the cumbersome armor of a medieval knight, be highly disadvantageous when maneuverability is required. In close quarters, under shelter, the boxfish's stubby fins provide sufficient movement. When flushed into open water, however, its potential speed is greatly reduced. On several occasions, I have seen boxfish and their triangular-bodied relatives, the trunkfish, with their caudal (tail) fins bitten off. Thus wounded, these inoffensive browsers lose all means of flight and tend to remain close to the shelter of the corals. The boxfish has traded maneuverability for armor plate capable of withstanding the attack of all but the largest predators, such as sharks.

Another form of this defense is found in the pufferfish, which trades maneuverability for inedibility. When it senses danger, it inflates itself with water to a huge, unmanageable girth. In its inflated state it is practically helpless, a huge distended ball capable only of rotating about its own center. Where the pufferfish and boxfish differ, of course, is that when the danger has passed the pufferfish expels its water and achieves some maneuverability. The prevalence of these pufferfish and boxfish in the sea is testimony to their adaptations for defense.

The Red Sea boasts several species of pufferfish, including the beguiling brown-and-white *Arothron stellatus* with its puppy-dog face and lustrous eyes, the ubiquitous porcupinefish (*Diodon hystrix*), and the white-spotted pufferfish (*Arothron meleagris*). The spiny globefish (*Arothron* sp.) is another species peculiar to this area. The spines of this puffer are so short that when its body is puffed, the spines form rather modest protrusions from the bulbous, inflated mass. The spiny globefish is a sedentary browser on mollusks and frequently can be found lying motionless on the sand next to a coral head. Thus exposed, this doe-eyed fish is often pounced upon and "massaged" by divers. Seeming to perceive its imminent role as dinner, the globefish hastily inflates itself to become as unmanageable a meal as its huge girth can create.

Among the most colorful predators on the Red Sea reefs are the groupers. The grouper *Cephalopholis argus*, for example, has a vivid red body covered with brilliant blue spots (Illustration 294). This bold fish is similar in many ways to its Caribbean counterpart, the coney (*Cephalopholis fulva*). When a diver approaches its coral head, the fish plays a rather elaborate game of hide and seek. It will utilize a maze of holes and crevices in the coral head, appearing first on one side, then the other, repeatedly ducking out of sight only to reappear moments later.

Another fascinating grouper is *Epinephelus fasciatus* (Illustrations 127–128). Normally this fish has a white body and a rust-red upper face. However, when it seeks shelter in an umbrella coral, it adopts a uniform rich orange body color. In this color scheme the fish is much harder to see, and will remain motionless until certain that it has been discovered. Then, as with so many fish, comes the last resort—explosively sudden flight.

A number of other groupers, from the giant jewfish (*Epinephelus gigas*) to the ornate and elusive lunar-tailed grouper (*Variola louti*), range the Red Sea reefs. The latter has a gaudy body coloration of blue spots on a red background and a yellow-edged inverted lunate tail with sharp tips.

One day we had an illuminating encounter with a small blue-spotted ray (*Dasyatis limma*). This ray, common throughout the Indo-Pacific, usually is extremely shy (Illustration 301). In most encounters it is found resting on the sand beneath a protective umbrella coral. When disturbed, it settles more firmly into its shelter and refuses to budge. Sometimes, however, if it is uncertain of the security of its refuge, it will bolt to some other, surer sanctuary. On this particular day we found a rather small ray, which we coaxed out of its shelter. To our amazement and delight the animal chose not to flee but began to play about us for several minutes. It even let itself be handled by our guide, doing loops and turns in a graceful ballet as it was manipulated by the diver's agile hands.

We had rather less rewarding experiences in trying to photograph the very large squid which frequent the coral shallows. These animals reach lengths of almost one meter. They seem to enjoy basking just under the calm surface of the inshore waters on brilliantly sunny days. Canny like all squid, these large creatures are extremely difficult to corner for photographic purposes. Moving forward or backward with great speed, they stayed out of our range with ease and left us thoroughly outmaneuvered.

Sharks abound in the Red Sea. The greatest concentrations seem to occur in the southern portion. Many are found in the areas near Port Sudan and are presumably plentiful all the way to the Indian Ocean and north to the Gulf of Suez. There seem to be fewer sharks in the Gulf of Aqaba, where one may dive for days with very few sightings. Of course, the sharks may well be there but not seen by divers due to other factors. The predators' well-developed vibration senses probe the human visitor as a potential food object and reject the diver long before visual contact is made.

One highly unusual inhabitant of the Red Sea reefs is the flathead (*Platycephalus*). Locally this large, sedentary fish—up to one meter long—is known as the "crocodilefish." One glance at the animal proves the cognomen apt. On a strongly tapered body, rather like that of the Caribbean lizardfish, sits a saurian head—long, wide jaws broaden to a flat cranium with vertical, protruding eyes. The eyes are well camouflaged by intricate fringes that extend from the upper and lower edges of the eye to cover nearly half its glossy surface. The crocodilefish sits on the open sand, motionless, awaiting a passing meal. When a careless fish swims within reach, the predator explodes from the bottom and snaps shut its huge jaws.

One of the most spectacular of the reef-dwellers in the Red Sea is the incredible poison-spined beauty known locally as the chickenfish and elsewhere by such names as lionfish, turkeyfish, zebrafish, and firefish. Two species—(*Pterois volitans*), and a relative known as the clearfin chickenfish (*Pterois radiata*)—are indigenous to many tropical Indo-Pacific reefs, but nowhere else in the world have I seen so many as here (Illustrations 289, 305).

The chickenfish are effectively and exquisitely armed with a brilliant array of long, poisonous spines on dorsal and pectoral fins. Venom is secreted in glands along the spines and carried to their tips in a groove under a thin sheath of external skin. The fish does not actively inject the venom, but when the spine penetrates the skin of prey or attacker the sheath is pushed back, leaving a smear of extremely potent venom in the wound.

Experiments on the effects of the venom have been conducted by Gerald Allen and William Eschmeyer, and, the results, as reported in the California Academy of Sciences Journal, *Pacific Discovery*, are unsettling. Injections of chickenfish venom in a variety of fishes invariably produced death within 10 to 30 minutes. Allen and Eschmeyer also reported an incident in which a man tried handling a chickenfish and was stung. Pain was immediate and became severe within five minutes. Within 35 minutes the victim was stuporous and in a state of shock, with no blood pressure and with heart registering only ten beats per minute. Epinephrine administered intravenously enabled the victim to recover. The swelling around the stings lasted for a month. Despite all this, the chickenfish are among the most sought-after subjects of the Red Sea; for, unlike many other specimens of the nightfeeding *Pterois*, they are quite active during the day.

When I have approached them, the chickenfish have risen slowly from their coral communities. Some hover just above their coral head, posing as if paid. Others are somewhat aggressive, lowering their heads and spreading their fins in a bull-like posture whose intent is unmistakable. Advancing on a diver, the chickenfish slowly drives the intruder back with the mere display of its formidable armament. I have always retreated, but grudgingly, and have managed to take a number of rewarding photographs. I have been informed that as many as 15 chickenfish will band together on occasion to drive divers from their territory with their dogged bulldozer tactics. Two of us, however, have managed to photograph as many as a dozen chickenfish in the shadows of a single coral head.

Sadly, these exquisite reefs of the Red Sea are not safe from destruction. Recent studies have shown that despite the most alert conservationist ethic (one of the areas studied has been set aside as a national undersea park, where fishing and coral collecting are forbidden), certain areas of these Red Sea reefs are being destroyed by chemical pollution. These reefs are near Eilat, Israel, but the conclusions would also be valid for much of the coastline of the Red Sea.

The study observed that periodic extraordinarily low tides exposed vast expanses of reefs to a fatal drying. Normally the processes of larval regeneration would restore these reefs within a period of about five years. However, this recovery process has been severely hampered because of pollution from both oil tanker operations and the loading of phosphate fertilizer on ships nearby. The assault is one that turns the coral polyp's own defenses against itself. Under normal conditions, coral responds to minor sand or other irritations by secreting ameliorative mucus. Bacteria in the mucus eventually decompose it and help in the recycling of nitrogen, phosphorus, and carbon. This entire system, in balance, aids in the coral reef's long-run survival.

Chemical pollutants, however, upset this mechanism. The original mucous secretion fails to halt the chemical irritation of the polyp. It continues to secrete mucus, which attracts bacteria that breed explosively. The coral is soon overwhelmed by the bacteria and covered with a green slime. Thus the normal recovery after the extreme tides fails to occur. All of this happens without even a highly publicized oil spill or phosphate "accident"—and in areas intended for preservation by the government and the populace. So while we celebrate the Red Sea for its magnificent undersea beauty, we hope man has not set in motion forces that will harm it irreparably.

Here, where the hot winds of the desert soar above the cold waters of the Red Sea, where ancient ruins such as those of Petra remind us of the fragility as well as the endurance of the human condition, our selective worldwide visits in the underwater wilderness come to an end.

280. *A lionfish (species possibly new to science) adopting a threatening posture against divers who have encroached on its territory. (Gulf of Aqaba, Red Sea)*

281. *A tiny but exquisite colonial denizen of the Mediterranean, the bryozoan* Repertora cellulosa. *(Alain Schweigert)*

282. *The Mediterranean demosponge* Acanthella acuta. *(Alain Schweigert)*

283. *Gorgonians of the Mediterranean. They look like a strawberry confection but are a soft coral. (F. Candela Ory)*

284. *Another colony of brightly-hued Mediterranean bryozoans. (F. Candela Ory)*

285. *A magnificently colored Mediterranean nudibranch* Peltodoris atromaculata. *(Alain Schweigert)*

286. *Spines of the Mediterranean sea urchin* Spaerechinus granularis. *(Alain Schweigert)*

287. *Blade-shaped coralline algae in the Mediterranean. These are plants, not invertebrates, which have evolved a crusty structure to discourage predation. (F. Candela Ory)*

283

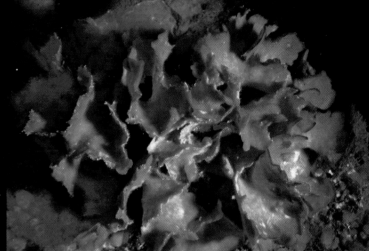
284

285

286

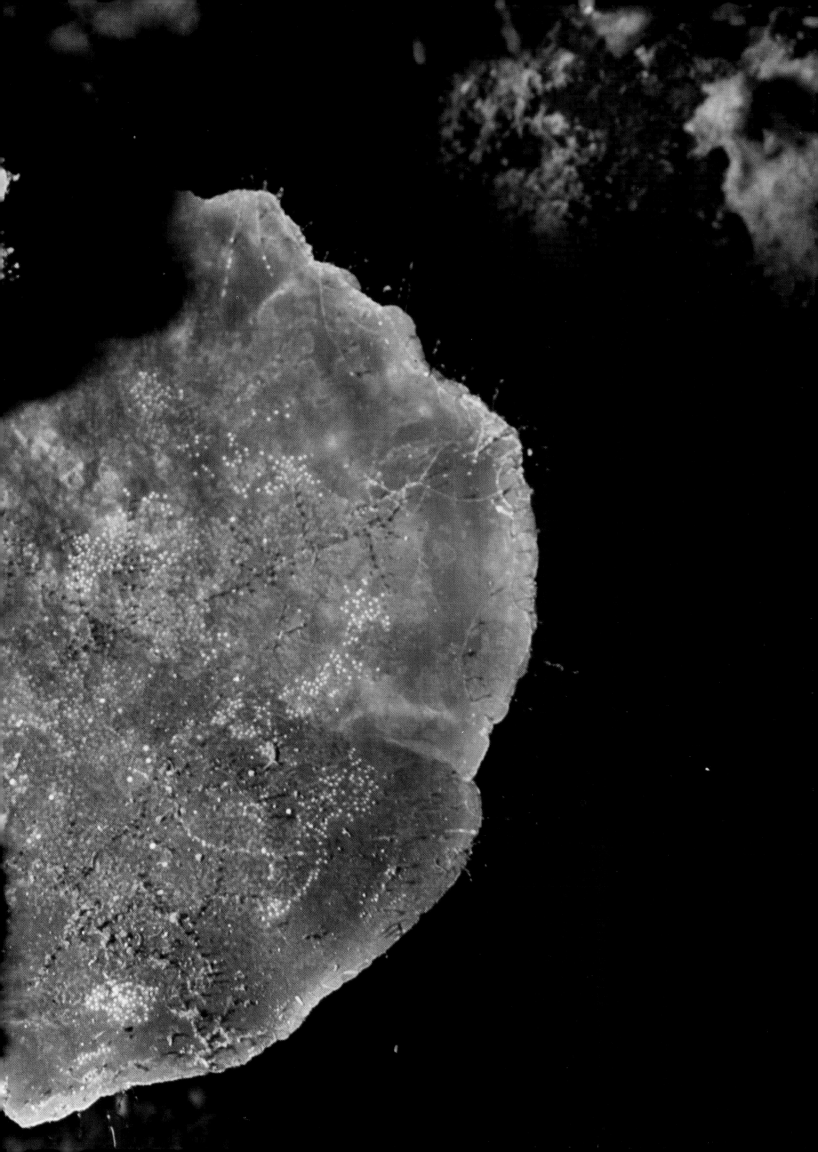

288. *An intricate small gorgonian growing on the surface of a boulder in the Mediterranean. (Alain Schweigert)*

289. Pterois volitans, *the lionfish, is also known as the turkeyfish, firefish, and zebrafish, among other names. (Gulf of Aqaba)*

290. *A nudibranch* Chromodon quadricolor *on a finger sponge. (Gulf of Aqaba)*

291. *The incredibly intense colors of* Dendronephthya *forming a vivid contrast with a stark background of coral skeleton.*

292. *A red finger sponge reaches upward from the coral like a hand. (Gulf of Aqaba)*

293. *Night-blooming coral* Tubastrea. (*Gulf of Aqaba*)

291

294. Cephalopholis argus, *a grouper, one of several that played hide-and-seek with the photographer, warily moving from one crevice to another.* (*Gulf of Aqaba*)

295. *Coloration of the male of the orange fairy basslet* (Anthias squammipinnis). *When a male is removed from its group, a female undergoes a color and sex change and takes its place.* (*Gulf of Aqaba*)

296. *Female* Anthias squammipinnis. (*Gulf of Aqaba*)

297. *The spectacular juvenile of the imperial angelfish* (Pomacanthus imperator). (*Gulf of Aqaba*)

298. Pomacanthus imperator, *adult imperial angelfish.* (*Gulf of Aqaba*)

299. Rhynchostracion nasus, *blue-spotted boxfish.* (*Gulf of Aqaba*)

300. Anthias squammipinnis, *orange fairy basslets, hovering above their protective coral head.* (*Gulf of Aqaba*)

301. *The blue-spotted ray* (Dasyatis limma). (*Gulf of Aqaba*)

302. *A colorful wrasse,* Thalassoma ruppelli, *moving swiftly among some coral heads.* (*Gulf of Aqaba*)

303. *The king angelfish* (Pygoplites diacanthus). (*Gulf of Aqaba*)

304. *A pair of red-tailed butterflyfish* (Chaetodon chrysurus) *browsing on coral polyps.* (*Gulf of Aqaba*)

305. *The lionfish* (Pterois radiata). (*Gulf of Aqaba*)

306. *A black jack* Caranx lugubris *in open water off the edge of an underwater precipice. It sweeps in to investigate the camera as two divers approach in the distance.* (*Cayman Islands, Caribbean*)

295

296

297

298

299

300

301

302

303

304

Conclusion: The Threat and the Promise

The underwater wilderness is in peril. An increasing number of marine scientists and others feel that the oceans of the world, and hence the wilderness areas within them, are dying or may die if current trends are not reversed.

Our generation has been set the task, on short notice, of somehow stopping a process, a calamity, that has been building for centuries. Oceans do not die overnight. Wasteful and destructive practices may be continued for centuries before the accumulation of massive populations makes their effects noticeable. The dying of the oceans is a very gradual process. However, it may well have developed such momentum that by the time human societies truly respond, it will be irreversible.

In ancient times native tribes were scattered and few in number. In the absence of modern techniques of preservation, they harvested the sea essentially for their own daily use. Sometimes their methods were haphazard and sometimes they required extraordinary skills, but they were never so efficient, or applied on such a scale, as to denude their reefs. If they did exert a harmful influence on any segment of their marine community, they could observe it directly and begin harvesting in some other area. These old customs were maintained until relatively recently by peoples in so-called "uncivilized" parts of the world. The coming of "civilized" man to these areas, however, had profound consequences for the underwater wildernesses and for the local tribes that used the sea. As we have seen time and again, native populations were subjugated and their resources taken by the conquerors. While the native populations had lived for eons by harvesting a small fraction of what their land and sea bore naturally, the conquerors, primarily Europeans, came from a radically different tradition. Their countries and governments had long been conquerors and exploiters. They invaded and took what they wished. Great quantities of minerals, pearls, copra, mahogany, tortoises, shells, fish, and even humans were loaded aboard the invaders' ships. Civilized man had brought a new philosophy: harvest without limit, and when the resource is exhausted, move on to new conquests.

The unrestricted plunder of resources by "civilized" nations was based on an amorality of remoteness. The ship captain in the tropics was the instrument of a large and powerful society with insatiable appetites for low-cost goods. The captain was merely the conduit by which these resources were transported home to the teeming populations whose agent he was. The man in the street in London or Boston was unaware that his country's whaling fleet was slaughtering rare tortoises in the Galápagos, or that one of its fishing fleets was inadvertently capturing and killing thousands of porpoises in great nets. By the same token, it was not important to the captains whether they killed off such animals. It was not their own country that they were harming. They simply did their jobs as they perceived them, according to the prevailing attitudes of their society. Much of this philosophy is still very much with us today. There are at least three different activities that should be questioned in light of the dangers they pose.

Commercial Fishing

Civilized man has always taken from the sea whatever kind of and however many fish he could eat or sell. The history of commercial fishing has been one in which advanced societies developed more and more sophisticated tools to catch ever-dwindling numbers of fish. Today, for example, tuna fleets prowl the world using sophisticated sonar devices and spotter airplanes to find their increasingly elusive and less abundant quarry. Eventually this persistence will drive the quarry populations to such low levels that they will not survive.

Sport Fishing

Throughout history the laurel of success in the hunt was the visible trophy. As societies became more city-bound, the need of men to prove their manhood with trophies has, if anything, become more pronounced. Fishing has always been one of the means to this end, and groupers, snappers, rays, marlins, and sharks have felt the fierce pressure of the trophy hunter. Tragically, as the trophy animal is hunted to extinction, the prestige value of each death is increased.

Souvenir Collecting

Perhaps the most ironic destruction in the underwater wilderness occurs when tropical reefs are pillaged of corals and shells to decorate coffee tables in civilized countries. Beauty is thus destroyed wholesale in celebration of beauty. The collectors ignore the inescapable truth that the coral they rip out is not easily replaced. On some popular reefs, no young coral stands have been left.

What has so far escaped civilized man's notice and concern is the desperately vulnerable state of the underwater wilderness. As we saw earlier, coral reefs are but a series of narrow bands of life along the eastern edges of the continents 20 degrees or less from the equator, and of scattered islands within the same tropical band. They are frighteningly finite. Can the underwater wilderness and the oceans of which they are a part really die? A hint of the answer may lie in a series of random observations, including some presented earlier in this work and others from recent news reports.

Illegal dynamite-fishing persists in Greece, Colombia, the Philippines, and other regions. Scientists estimate that only 10 percent of the fish killed in an explosion rise to the surface for collection; 90 percent simply sink to the bottom and decompose. In the 1960s I went to the reefs in the Netherlands Antilles with friends who boasted that in earlier years they each could spear a dozen or more fish per dive. Now the same sites are empty of food fish. My friends still dive with their spearguns and wonder why the big fish are gone. Ironically, the native population here took on the destructive habits of their conquerors but, unlike them, cannot move on to fresh new territories.

Hundreds of kilometers of the Great Barrier Reef have undergone severe damage from crown-of-thorns starfish and other causes. The reasons for the destruction as well as possible steps to contain it remain unknown, but scientists suspect that man's pollution and souvenir collecting of shells were at least contributory factors.

Penguins in the Antarctic are being poisoned by pesticides from Africa.

In the Red Sea, coral regeneration is hampered by low-level pollution.

In Baja California and in the United States, the Mexican and American governments have moved to ban Japanese and other foreign fishing boats because of sharply decreasing fish catches.

In Maine, the lobster industry slowly shrinks as that prized crustacean population dwindles.

The Connecticut oyster industry struggles back from near extermination with government-supported programs to create new oyster beds in unpolluted waters.

In Hawaii, the price of puka shells used in necklaces has escalated as supplies run short.

In Ecuador, American tuna boats working the Humboldt Current are impounded because they compete with the Ecuadorian fleets and catches are short.

Honduran shrimp fishermen are faced with competition from Mexican shrimp fleets and by a dwindling supply.

Bitter American fishermen in California blame their declining catches on Russian trawlers working off the coast.

Tuna fleets now pursue schools far into the South Atlantic and the south of Australia with sonar and spotter planes. Many of the traditional tuna grounds have been swept clean of these sleek ocean roamers.

One species of whale after another heads toward extinction. However, the whaling nations cannot bring themselves to voluntarily end the slaughter. It seems that only when there are no more whales to kill will finis be written to this hunt.

In the Antilles, fishermen drop fish traps onto the richest reefs. Fish are attracted to the baited cages and trapped almost immediately. Divers note that the traps are left untended for several days; and in that time dozens of imprisoned triggerfish, butterflyfish, moray eels, and other species starve to death. Most of these decomposed, dead animals are simply thrown back into the sea.

In Palau, Micronesia, plans have been drafted for a massive superport, thus threatening the pristine far Pacific with the same dangers now facing the Red Sea reefs.

The list could go on and on, but the message is clear. Man's past relationship with the sea has been thoughtless in the way a small child is thoughtless in its treatment of a pet: the harm is done without malice, but it is done nevertheless. The result has been destruction beyond measure. In our generation it seems as if the numbers of mankind whose wastes poison the sea, and yet who must feed from the sea, are finally all the sea can bear. Now it has begun to fail us.

Is there hope for the underwater wilderness? While some say no, others point to developments that could turn the tide of destruction. Certain of these could even allow us to reap great benefit from the sea without inflicting harm on it.

In places as far apart as England and Oregon, large rivers have begun to support new fish populations after being poisoned by effluents for a generation. It has become clear from these examples that when the ravages of pollution begin to affect enough consumers directly, pressures quickly mount to end the pollution. This is true even when the solutions involve some sacrifice by those consumers.

Similarly, new technologies for cleaning up oil spills, for purifying water, and for reducing effluents have been developed largely when crises caused an aroused public to demand and pay for them.

The sea and its animals and plants have remarkable resiliency and great recuperative powers. Given even an outside chance—a slim margin of survival through lessened pressure—and many populations rebound dramatically. The California sea otter, reduced in the 1940s to perhaps a few hundred individuals, was considered doomed. Scientists felt that such a small and pressured community could not replace its dwindling members. Strong prohibitions were placed on killing the otters, and by the 1970s colonies of the frisky animals were spreading up and down the Pacific coast.

The survival of the underwater wilderness will be made possible only when it is universally realized that the oceans and reefs of the earth are incredibly complex societies and not only a critical source of food for mankind. The societies are composed of unimagined numbers of individuals whose feats of chemistry and body structure rival and frequently surpass anything found on land. However, these amazing societies and the creatures who comprise them exist only within certain restrictive conditions. They are extraordinarily vulnerable to man's actions. We must awaken to the unavoidable fact that the same processes that smog our air, poison our rivers and lakes, and snuff out entire species could eventually destroy the underwater wilderness.

It would be disastrous for man to allow the underwater wilderness to die. We have observed earlier that everything seems to be interconnected in the sea. The results of harm done at one point can be unexpectedly felt many thousands of kilometers away. If the still-healthy regions are lost, we will have witnessed a watershed in processes whose momentum may well go on to destroy the oceans themselves. If the oceans subsequently die, man as the end consumer of their food and algae-generated oxygen must perish as well.

The alternative is easy to state and agonizingly difficult to put into practice. We must learn to live in harmony with the sea and its society. We must fish within an international plan of resource management. No species should knowingly be hunted to extinction. If any species fails to replenish itself, hunting pressure must be reduced.

The slaughter of whales, fur seals, groupers, and tortoises until they are nearly exterminated must be stopped. Such practices must become footnotes to a barbaric history whose like we should never again allow. We can harvest from the sea all it can continually replenish—that, but no more. We must stop emptying into the sea the concentrated poisons of our industries.

If we succeed in formulating resource management plans and in effectively disciplining those who utilize the sea, we may have met one of our century's great tests: to pass on to our inheritors all that is precious in the underwater wilderness.

Index

319